Geography and Gender

In the 1980s feminist geography offered a stimulating new approach to the subject, providing fresh perspectives on traditional areas of the discipline. Originally published in 1984, the authors of *Geography and Gender*, members of the Women and Geography Study Group of the Institute of British Geographers, compiled an introductory text with an accessible, concise and jargon-free style which engaged the reader's interest. It challenged the current thinking about geographic research and teaching at the time and suggested important new directions.

The book opens with an introduction to feminist geography. It then demonstrates how a feminist approach changes and improves our understanding of geographic processes and patterns. Urban development and structure, industrial location and spatial variations in employment, access to facilities and processes of the developing world are re-examined from a feminist perspective. Finally, it discusses how a feminist approach can change the ways in which both teaching and research in geography are carried out.

Interest and research in feminist geography had been growing both in Britain and elsewhere, but until now no introductory text had been available. It will also be useful to anyone wanting a concise and authoritative introduction to feminist ideas and their application in the area of geography and planning. Each chapter includes an annotated further reading list and topics for discussion. Suggestions for individual or group project work, keyed to topics discussed in the text, are given in the final section.

Geography and Gender

An Introduction to Feminist Geography

Women and Geography Study Group of the IBG

Routledge
Taylor & Francis Group

First published in 1984
by Hutchinson & Co. (Publishers) Ltd

This edition first published in 2025 by Routledge
4 Park Square, Milton Park, Abingdon, Oxon, OX14 4RN

and by Routledge
605 Third Avenue, New York, NY 10017

Routledge is an imprint of the Taylor & Francis Group, an informa business

Publisher's Note
The publisher has gone to great lengths to ensure the quality of this reprint but points out that some imperfections in the original copies may be apparent.

Disclaimer
The publisher has made every effort to trace copyright holders and welcomes correspondence from those they have been unable to contact.

A Library of Congress record exists under ISBN: 0091566711

ISBN: 978-1-032-89339-6 (hbk)
ISBN: 978-1-003-54258-2 (ebk)
ISBN: 978-1-032-89399-0 (pbk)

Book DOI 10.4324/9781003542582

Geography and Gender

An introduction to feminist geography

Women and Geography Study Group
of the IBG

Hutchinson
in association with
The Explorations in Feminism Collective

London Melbourne Sydney Auckland Johannesburg

Hutchinson & Co. (Publishers) Ltd

An imprint of the Hutchinson Publishing Group

17–21 Conway Street, London W1P 6JD
and 51 Washington Street, Dover, New Hampshire
03820, USA

Hutchinson Group (Australia) Pty Ltd
PO Box 496, 16–22 Church Street, Hawthorne,
Melbourne, Victoria 3122

Hutchinson Group (NZ) Ltd
32–34 View Road, PO Box 40–086, Glenfield, Auckland
10

Hutchinson Group (SA) (Pty) Ltd
PO Box 337, Bergvlei 2012, South Africa

First Published 1984

Set in 10/11 point Plantin by Saxon Print, Derby

Printed and bound in Great Britain by
Anchor Brendon Ltd, Tiptree, Essex

British Library Cataloguing in Publication Data

Institute of British Geographers. (*Women and Geography Study Group*)
Geography and gender.—(Explorations in feminism)
1. Geography
I. Title II. Series
910 G116

Library of Congress Cataloging in Publication Data

Geography and Gender.

Bibliography: p.
Includes index.
1. Women geographers. 2. Geography—study and teaching. 3. Feminism.
I. Women and Geography Study Group of the IBG.
II. Explorations in Feminism Collective (Great Britain)
G65.G46 1984 910′.88042 84–12871

ISBN 0 09 156671 1

For all the women geographers who went before us. Although often unacknowledged at the time, they have made our way easier

Contents

Figures

Tables

Acknowledgments

The authors and publishers would like to thank the copyright holders below for their kind permission to reproduce the following material:

Figure 3 copyright Harlow Council, Information Services Dept, Harlow.
Figure 7 Ministry of Information, People's Democratic Republic of Yemen.

Foreword

Feminism is clearly on the agenda in geography. In January 1984 the session of the Women and Geography Study Group was one of the best attended and liveliest at the Annual Conference of British Geographers. For a Group which had constituted itself only a few years before, it was a dramatic success.

It was also, as a number of people have remarked, both then and subsequently, about time too. Feminist geography had been preceded by, and in many senses had had the ground prepared for it, by other critiques of established modes of thought. Both radical/marxist and humanist critiques opened up fertile ground for feminist analysis. But what feminist analysis has done is not merely add another dimension. It has in a very important sense been instrumental in pushing both critiques further and in new directions.

Take the case of the geography of employment. In recent decades there have been enormous upheavals in the gender composition of the waged workforce, and these changes have brought home many previously unexamined interrelations: between waged labour and domestic labour, between the organization of production and the organization of the family, more generally between what is customarily called the economic and the social, even between industrial and social geography. But if we go so far, we should also go further and recognize, more than is commonly done, the interpenetration of the economic, the social and the ideological, in the processes we study and the categories we use. One of the things feminist geographers are beginning to throw into high relief is that many of the apparently simple economic processes we study just aren't so simply 'economic' after all. In the past, industrial geographers have taken labour-demand as given – that in electronics, for instance, men do the R & D and women do the assembly. From a geographical point of view such things matter: were it not for that sexual division of labour in production there would not be that geography of employment. In the 1970s it was an advance for radical analyses to point out that the nature of labour requirements within industry should not be taken as given but had something to do with the process of production, and that changes in location were often, thereby, related to changes in the labour process. Too often, however, this was where the analysis stopped: nature of labour process leads to type of labour-demand, leads to locational pattern. But things are not so simple. The nature of the labour process does *not* uniquely determine the nature of labour-demand (*why* men doing R & D? *why* women doing assembly?) and not, therefore, location. In many cases the nature of labour-demand has very little to do with the 'technical' requirements of production. It is more a social question. There is another process which intervenes – that whole complex of social causes through which different parts of the population are allocated to different kinds of jobs within society as a whole. And in that complex of causes, gender relations are of central significance. And, too, within British geography, it has been in feminist analyses that that point has been most clearly and most forcefully made.

That's one example; there are many more. It was from a feminist perspective that our accepted categories of 'skill' came in for some of the most devastating criticism. Again, without some grasp of the gendered-construction of the British labour market it is quite impossible to understand the recent locational contortions of British capital as it has tried to restructure, and relocate, itself out of crisis. A feminist perspective, then, in the widest sense, has a lot to contribute to better explanation within geography.

And geography, too, has something to contribute to feminism. It is strange that feminist analyses to date have largely ignored spatial variation, certainly within individual countries. Perhaps this was because, initially, there was quite enough work to do in getting to grips with major, general, mechanisms. Maybe, too, there was a political desire to treat 'women' as a single coherent category. But gender roles and the lives of women vary quite dramatically even within so small and comparatively homogeneous a country as the UK. Unearthing and understanding these differences adds immeasurably to our analysis of the way in which male dominance works – and of the remarkable versatility of patriarchy.

There is a long way to go. Much feminist geography so far has focused on women, and in some ways that is a reflection of the stage we are at – still establishing the issue, and even the category, as significant to analysis. But feminist geography is also more than studying women. A feminist analysis in the broadest sense is one done from an anti-sexist perspective, and can be concerned with any subject. Men are gendered too. It is just as much a question why men dominate R & D, as why women do the assembly.

There is a question, too, of the way in which we work. The language of geography is often still remarkably archaic. We all know the patiently-pained expression from across the table at the meeting when we've just, hesitatingly or ironically, interjected '. . . or she . . .'. The look which says 'OK, OK, do you *have* to point it out *every* time? . . . (Why not: we are excluded every time?). . . You *know* when I say "he" I include women'. Such breathtakingly inadequate conceptualization would be jumped on in any other context. And how many courses are still on offer in geography departments up and down the country with the almost comically narrow concern of 'Man and the Environment'? Such language is more than a bad habit. It is symptomatic of far deeper levels of sloppy thinking.

So there is a way to go. The chapters in this book represent a significant step forward – an important first landmark – on what is likely to be a long, difficult, but fascinating, road.

Doreen Massey
Professor of Geography at the Open University

Preface

This book is a new departure, in its content and in the way in which it was written. It is the first undergraduate text on feminist geography and it has its origins in meetings held on the fringes and as part of the Institute of British Geographers' annual conferences over the last few years. At these meetings a number of geographers – not all women – got together to provide a forum for exchanging views on feminist teaching and research. We soon realized that there was an urgent need for a textbook: to meet a need in our own work and to help us with the myriad of requests for help, advice and reading lists from our colleagues and students at many institutions. This book is the result. It was written co-operatively by nine members of the Women and Geography Study Group, which was formally constituted in 1982. They are, in alphabetical order, Sophia Bowlby, Jo Foord, Eleonore Kofman, Jane Lethbridge, Jane Lewis, Linda McDowell, Janet Momsen, John Silk and Jacqueline Tivers. The text was typed by Angela Jordan, Roger Sherwin and Colin Watson, and figures were drawn by Sheila Dance. Peter Pearson helped with the index. Sophia Bowlby, Linda McDowell and Jacqueline Tivers undertook the organization, co-ordination and editing that such a co-operative adventure of necessity entailed. We decided not to attribute the chapters to individuals as we all, individually and in groups, discussed drafts, made amendments, and agreed the final version.

It was an enjoyable experience and we would like to record our thanks to all members of the Group who, in many ways, helped us to clarify the ideas in this book. We hope we've provided a challenge to current practice and a stimulus to change.

PART I
An introduction to feminist geography

1 Why study feminist geography?

A 'geography of women' or 'feminist geography'?

Looking through a representative selection of geography books on a library or bookshop shelf, it appears as if most geography is concerned with 'man'. We are confronted by 'man and his physical environment', 'man and culture' or simply 'man and environment'. The authors of such texts may not intend to portray humanity as being entirely *male*. Nevertheless, it is this image which is created in the mind of the reader and which persists when the subject matter of the books is consulted. We are presented, for example, with *man* as the agent of change in agricultural landscapes, *men* digging for coal (or being made redundant by the closure of coal-mines), and the results of surveys in which *men*, as heads of households, have been asked for their opinions on recreation resources, transport needs or housing. We might, in fact, be forgiven for thinking that *women* simply do not exist in the spatial world.

If we turn to the historical study of human activities, we find the same situation. One of the features of human history appears to be that groups which are disadvantaged, being in some way dominated or oppressed by other groups – slaves, the common people, non-whites, women – do not figure as prominently in historical accounts as their social significance or sheer weight of numbers would suggest. Often, such groups are barely mentioned at all. In the majority of historical studies, the researcher is concerned with the way in which *men* gradually increase their control over nature, with the exploits of warriors, princes and kings and, more recently, of various states*men*, business moguls, inventors and scientists. The history of dominated groups has been taken as more or less incidental to the lives of those characters who, although few in number, occupy the centre of the historical stage and exercise control over society's wealth and resources. If the activities or experiences of subordinate groups are considered, they are often portrayed as part of some natural, even divinely-inspired, order, represented in such a way as to suggest that members of such groups are happy with their lot, or they are simply denigrated.

Rowbotham (1979), challenging such distortions, has suggested that women are 'Hidden from History'. We contend that women are also 'hidden from geography'. Hayford (1974) clearly makes this point when examining the situation in geography in North America. She argues that, although geographers recognize the existence of women, they make little effort to investigate the role they play in society other than in terms of adjustment to a male-dominated and male-determined order. Later in the 1970s Tivers (1978) published an article suggesting that geographers should consider 'How the other half [i.e. women] lives', and drop their patronizing attitude, common to workers in the other social sciences, to the study of women's perceptions and behaviour. During the late 1970s and early 1980s, a variety of work has appeared in which women's spatial behaviour and environmental perceptions have been highlighted, and the papers by Monk and Hanson (1982) and by Zelinsky, Monk and Hanson (1982) provide numerous examples in many fields of geography. This work has quite clearly demonstrated the existence and effects of women's inferior economic and social status, and shown how traditional geographical work is deficient in this respect.

Despite the increase of interest shown by some (although still not many) researchers within geography in women's activities and problems, the writers of introductory texts seem to have remained largely unaffected by such a change of attitude. One of the aims of this book is, therefore, to redress the balance of introductory geographical material in favour of *women*. Chapter 3, therefore, brings to our attention how changes in urban spatial structure reflect and reinforce women's social position, while Chapter 4 is concerned with the reorganization and relocation of industry and employment in post-war Britain as these relate to women. Chapter 5 deals with the problem of women's access to facilities, looking in particular at the provision of health care. Chapter 6 considers the role of women in the social and productive organization of Third World countries. Finally we discuss problems relating to methods of teaching and research in geography as a whole in Chapter 7, along with proposals for alternatives.

Encouraging geographers to consider women's daily lives and problems as legitimate, sensible and important areas for research and teaching is, however, not the principal aim of this book. We are concerned that there should be a geography of women as well as of men. Nevertheless, we do not wish this to be seen as merely 'adding' women to existing types of geographical analyses without any alteration to the theoretical assumptions underlying these analyses. Making women visible is simply not enough. The very fact that differences between male and female spatial behaviour patterns can be so readily identified does not in itself guarantee that the geographer will do anything other than conclude: 'Well, men and women are different, and it's interesting to see how this comes out in their behaviour'! In common with other approaches in geography which are critical of mainstream work, we must analyse and understand why women remain in a subordinate position.

What we argue for in this book is not, therefore, an increase in the number of studies of women *per se* in geography, but an entirely different approach to geography as a whole. Consequently we consider that the implications of *gender* in the study of geography are at least as important as the implications of any other social or economic factor which transforms society and space. We use the term 'gender' to refer to *socially created* distinctions between femininity and masculinity, while the term 'sex' is used to refer to biological differences between men and women. Hence, we are concerned to introduce the idea of *feminist geography* – a geography which explicitly takes into account the socially created gender structure of society; and in which a commitment both towards the alleviation of gender inequality in the short term and towards its removal, through social change towards real equality, in the longer term, is expressed. The question of the relationship between feminism and geography, and its implications for feminist geography, is raised in Chapter 2. The succeeding chapters then serve to illustrate how a feminist approach alters the nature of various traditional 'geographical' topics. The book is not, however, intended to provide a comprehensive coverage of all issues relating to feminist geography – either in theoretical background or empirical examples – but rather is intended as an introductory text, in which essential concepts are outlined and explained. The reader is encouraged to undertake the task of further research and applied studies through following up the project proposals suggested at the end of Chapter 7. These are intended for use either by undergraduates working on group projects or as undergraduate dissertation topics.

Why bother?

So far we have outlined the general approach of this book. But, before any further consideration of feminist geography, we should answer one very obvious question which may well be in the mind of the reader – why bother?

Why should we, as geographers, be worried about sexism or the 'invisibility' of women? Surely things are different these days and women have equal rights with men? Why should we any longer need to stress the existence of inequality based on gender?

To take the last point first, it is perfectly clear that inequalities between the sexes have not been eradicated within advanced capitalist societies. For example, it is true that in Britain, as in many other similar countries, a certain amount of *legislative* change occurred in the 1970s. The Equal Pay Act was introduced to prevent the payment of unequal wages to men and women for doing the same or similar work. However, employers have found many ways to evade the legislation, and their task was made easier by a five year easing-in period. It is extremely difficult to compare the wages paid to men and women for similar work because, in so many instances, workforces have been increasingly segregated so that the scope for job comparison is limited. The Sex Discrimination Act which was intended to provide equal opportunities and which covers education, training, employment, housing and the provision to the public of goods, facilities and services, appears far-reaching, but is extremely complex and full of loopholes. In 1977, the Employment Protection Act also came into force, giving women the legal right, given two years' full-time service, to return to their job after having a baby. Women were also granted the right to maternity pay under certain conditions. Such measures do represent a small step towards providing a degree of financial independence for women. However, two areas which have as yet remained resistant to legislative change but in which men and women are treated unequally, are the tax and social security systems.

However, legislative change does not necessarily herald a fundamental change in societal organization or norms and values. This has been the case partly because of the very limited nature of the legislation. For example, while 'equal pay' and 'equal opportunity' have been officially supported as worthy *concepts* to be worked towards, no attention has been paid to the practical need of working mothers for child care, in order to allow them to work in the first place. In addition, even the limited though positive gains made on behalf of women during the 1970s have begun to be eroded as the current economic recession has deepened. For example, it is now much more difficult for working women to qualify for maternity pay, and the Conservative government attempted (although unsuccessfully) in 1980 to 'rationalize' various forms of maternity benefit, to the detriment of working women. In many ways, this situation is not surprising. Someone has to foot the bill for child care if women no longer provide it, for the increased wages (if women truly obtain equal pay and opportunities), and for a whole range of maternity benefits. The money can be raised directly from private or public employers if they provide free child care and other benefits for their employees, or indirectly via various forms of taxation. During times of economic prosperity, profits are higher and 'society' can more readily afford to accede to at least some of the demands made by women (bearing in mind the fact that many other groups will also be struggling to obtain 'a larger share of the cake'). During times of recession, not only is there likely to be a squeeze on wages in many sectors of the economy, but also public expenditure is more likely to be cut. Both these trends have been only too evident in Britain since the late 1970s.

Inequality of employment opportunity is not the only area in which women remain subordinate to men. Of particular importance is the difference in educational attainment between men and women. At all levels above GCE 'O'-level men predominate. There are more male graduates, and therefore more men than women entering managerial and professional occupations. Women are clearly *not* inferior to men in terms of intelligence or application to study. As in the case of employment, there are social factors which intervene to prevent women from achieving their full potential in many cases. In an interesting article by Linda McDowell (1979) on the gender structure of British geography departments, statistics were presented to show that, whereas 42 per cent of full-time student geographers were women, this was true of only 7 per cent of full-time university geography teachers. However, 44 per cent of part-time teachers were women. Part-time employment, offering lower status and less security than full-time service, is the most 'convenient' way to utilize the talents of women, without at the same time disturbing the societal 'status quo' to any significant extent.

The vast preponderance of full-time male academics in geography departments in institutions of further and higher education, therefore, both reflects and perpetuates a pattern of social relationships which are male dominated. This brings us back to the first of our original points – why should *geographers* be concerned about the gender structure of society? The answer to this question is not only political. In recent years geographers have been active in research into questions of societal inequality based on class, income or race, not only because of a growing belief in the need to expose injustice and work towards a better and fairer future, but also because of the importance of an accurate analysis of human geographical phenomena for the development of our understanding of society and space. We cannot interpret human behaviour unless we take into account all the societal patterns and structures which exist. The same is clearly true of a feminist perspective in geography. Without an understanding of the gender roles which underlie the workings of society, we cannot hope to present a reasonable analysis of the spatial behaviour of women and men, nor of the institutions both dependent on and influencing that behaviour.

It is because of our belief in the importance of a feminist approach to geography that the Women and Geography Study Group of the Institute of British Geographers has been formed, and, through this book, seeks to put forward an introductory explanation of such an approach.

2 Feminism and geography: theory and practice

Introduction

In this chapter, we look first at the main features of feminism and introduce some of the most important concepts in feminist theory and practice. In particular, we examine the notion that people form distinct *classes* in our society, the associated concept of *patriarchy*, and the idea that women's lives differ from men's because so many women have *dual roles* as both domestic and waged workers. We shall show how important such concepts are if geographers are to understand fully both the decisions taken at various levels about the location of homes, workplaces and public and private facilities and also the interrelations between these decisions and the differing day to day behaviour, experiences and opportunities of women, men and children. Finally we examine links between feminism and other approaches in geography.

Feminism and feminist ideas

One of the key concerns of feminism has been to draw attention to the contrast between the lives of women, centred upon the 'personal' or 'private' sphere of home, family and domestic concerns, and the lives of men which are centred upon the 'public' sphere of waged work and formal political activity. The identification of these two spheres with women and men respectively generally tends to be interpreted as meaning that men and women are equal but different. However, a feature of all feminist argument is that these differences in the social position of men and women systematically work to the advantage of men so that women and men in fact have unequal power, opportunities and social prestige. Thus, although, as we shall see, feminists differ among themselves in the relative importance accorded to inequalities deriving from differences of class and race, they agree that relations of unequal power between men and women are an extremely important feature of society, and the reason for women's inferior social position.

A further, related, feature of feminist approaches is that they question whether what are often thought to be *innately* 'masculine' or 'feminine' characteristics (such as 'logical thought' and 'toughness' in men, and 'intuition' and 'gentleness' in women) are socially rather than biologically determined. Thus, as we have already seen in Chapter 1, feminists draw a distinction between the terms sex and gender. They use 'sex' to refer to the biological differences, associated with reproduction, between men and women and 'gender' to refer to socially created distinctions.

Feminists therefore examine differences between the power, social position, attitudes and behaviour of men and women, put forward theories to account for them and, most importantly, explore ways in which current practices in society might be changed in order to release women from their subordination. An immediate implication for feminist geography is that, while it may be necessary, it is not sufficient to examine men's and women's spatial behaviour patterns and perceptions of space, find that they differ, and document the differences. At its worst, this can simply lead to a proliferation of topics in (or branches of) the discipline in which the spotlight is simply turned on to women: as in, for example, 'women and development', 'women and transport geography', 'women and housing'. . . . At its best, such work may document and demonstrate the social inequalities facing women, but it will not explain *why* these have arisen. Rather, two alternative approaches have been advocated by feminist writers. The first examines women's and men's behaviour in terms of the *interrelationships* between the two genders, setting aside other fundamental cleavages in society. This kind of approach is characteristic of the approach of *radical feminists*, who draw heavily on the concept of *patriarchy*, which is explained later. The second approach, advocated by *socialist feminists*, attempts to link gender relations to the wider framework of social relations which exist in society as a whole, and which is structured by other factors as well as gender differences. In Britain – an advanced, capitalist society – class and race are major cleavages which interact with gender to produce complex patterns of dominance and subordination. If factors such as religious or regional differences are also included, the picture becomes extremely complex. In Ireland, for example, the religious differences which are intimately bound up with the struggle between republican nationalism and unionism have strong implications for women's rights on contraception and abortion. Women fighting for causes such as civil rights in North America, or for socialism or Irish nationalism in Britain, are too often expected to subordinate their own needs and defer their own struggle. The interrelations between gender and class have necessarily attracted the attention of socialist feminists, with some consideration also given to those involving gender, class and race (e.g. Davis, 1982).

We now consider the main features of radical feminism and socialist feminism.

Radical feminism

Radical feminists regard the subordination of women by men as the fundamental inequality in all human societies. In *The Dialectic Of Sex*, Firestone (1970) argues that a division of labour between the sexes preceded and gave birth to the division of labour by class and race. She further argued that the elimination of sexual oppression should lead to the elimination of other oppressions. The primary cause of conflict in society is located in the struggle between men and women over the social relations involved in *biological reproduction*. Thus there are two fundamental classes in society, consisting of women and men, rather than classes such as feudal lords and serfs or capitalists and workers. From the dawn of history, Firestone argues, pregnancy and the direct dependence of small children upon their mothers for food have put women at a disadvantage and so enabled men to gain an upper hand. Although it is now possible for women to control their fertility, and for children's needs to be provided for without the constant presence of the mother (at least in certain parts of the world), men have managed to keep their dominant position by perpetuating ideas and practices which ensure the continued dependence of children upon adults and of women upon men. Just as Marx and Engels urged the working classes to seize control over the means of economic reproduction, so Firestone argues that women must seize control over the means of biological reproduction, in order to free themselves from male oppression. However, many radical feminists argue that it is the social, cultural and economic structures which surround the biological difference between men and women that must be undermined, rather than the biological difference itself – in short, the emphasis must be placed on *gender* roles rather than *sexual* roles.

The concept of *patriarchy* has been used to explain why sex (a biological fact) becomes gender (a social phenomenon). In *Sexual Politics*, a book which deals with the United States but is relevant to other advanced capitalist societies, like Britain, Kate Millett (1971, p. 25) states:

> Our society . . . is a patriarchy. The fact is evident at once if one recalls that the military, industry, technology, universities, science, political offices, finances – in short, every avenue of power within the society including the coercive force of the police, is entirely in male hands.

Patriarchy can be defined as a set of social relations between men which, although hierarchical, establishes an interdependence and solidarity between them which allows them to dominate women. Thus, although men of different ages, races and classes occupy different places in the patriarchal pecking order, they are united because they share a relationship of dominance over their women. To some extent, patriarchy works because men at upper levels in the hierarchy can 'buy off' those at lower levels by offering them control over at least some women (Hartmann, 1981).

It is argued that the basis of patriarchal power lies in men's ability to control the kinds of work that women do. In a society such as our own this control is not only a matter of immediate physical force, but is also exercised through men's control of economic organization and social institutions, and through the perpetuation of attitudes and beliefs which justify men's dominance in society. It is important to stress that the essence of such *ideological* dominance is that ideas favourable to male power are accepted by both men *and* women. Moreover, acceptance of ideas of women's inferiority (for example, that women can't understand machinery) often lead to patterns of behaviour by both women and men which appear to 'prove' that inferiority (for example, women don't learn how cars work, men do, therefore women know less about cars than men).

Radical feminists argue that through the exercise of male power, women's work is largely confined to the 'private sphere', with the result that they carry out many (unreciprocated) personal and sexual services for men as well as those of child-rearing. They stress that women's sexuality and control over their own bodies are restricted by men in a way that is rarely true of men by women. For example, in Britain if women need medical services for contraception, abortion or ante- or post-natal care, they almost invariably find such services controlled by men (see Chapter 5). Furthermore, women are systematically prevented by men, as employees, fellow workers and husbands and fathers, from obtaining jobs that pay a decent living wage. Discriminatory wages, hiring and work practices and domestic demands limit women's employment prospects (see Chapter 4). Assumptions that women are unable to cope with jobs demanding leadership, independence and toughness of mind or body are often used as justifications for paying women low wages and not offering them promotion, as is the assumption that women are 'naturally' likely to interrupt their working lives to care for children or elderly relatives. As we have suggested, such assumptions are frequently self-fulfilling prophecies. Despite the legislation in Britain during the 1970s for equal pay, against sex discrimination and to ensure a woman's right to maternity leave, there is considerable scope for evasion on the part of employers as we have pointed out in Chapter 1. Since so many jobs are segregated by gender, comparison between a 'man's rate' and a 'woman's rate' for the same job is often difficult if not impossible. So most women still do lower paid 'women's work', often on a part-time basis in order to cope with the demands of their domestic responsibilities.

Patriarchal relations, therefore, work to ensure that the vast majority of women are legally and economically dependent on men. Thus, because of the restrictions placed upon them, women tend to live through their husbands or partners and through their children, rather than for themselves. Women confined to domestic work are often thwarted because many outlets for their energies and talents are closed and they are frequently isolated and bored (though rarely idle!). More commonly, a woman has to fulfil a *dual role*, combining poorly paid, part-time, 'proper' (i.e. waged) work with domestic work. Not only is women's work in the private sphere generally denigrated but also their waged work is often described in derogatory terms such as being 'only for pin-money'. Finally, women are liable to all kinds of verbal and physical harassment from men, both at work and in public places and, through domestic violence, in their homes. This 'sexist' behaviour on the part of men plays an important role in 'keeping women in their place' and thus perpetuating men's control over the behaviour of women.

The foregoing discussion has emphasized that male domination can operate at several different levels – for example, through male control of organizations, institutions and their rules, and also, at the personal level, through family, work and social relationships. For individual women, the results of this domination are felt in their everyday lives as a limitation of both their immediate and long-term opportunities. This suggests that feminist research in geography needs to be concerned with how this limitation of opportunities is experienced, paying particular attention to women's relationships with men and other family members, both in the domestic sphere and at work. For example, the availability and price of public and private transport, together with the number, size and spatial distribution of facilities used by women in their domestic role are clearly of importance in influencing women's opportunities.

Many of the insights provided by 'time geography' (e.g. Thrift, 1976) are valuable in analysing women's opportunities because constraints on spatial behaviour are considered in terms of *all* the tasks that an individual wishes to fulfil, and questions of timing, constraints (e.g. opening hours of facilities), and 'authority constraints' (e.g. patriarchal assumptions and practices affecting the availability of resources to women), may, in principle, be considered. Most housewives have quite complex schedules to fulfil, and for the majority of women who do waged work, problems of timing and co-ordination may be even more critical, especially when child care must also be found to cover those periods when a woman is at work. Even assuming no fundamental change in the current division of domestic and waged labour, considerable scope exists for improving women's lives by rearranging the spatial and temporal pattern of availability of facilities, as discussed in Chapter 5, pp.90–3. If major changes in gender roles are assumed, the implications for changes in the built environment may be considerable – those radical feminists who think liberation possible only if the two genders are completely spatially segregated, represent the most extreme position. Foord and Mackenzie (1981) have examined some of the implications of such views, while other experiments with, and visions of, a non-sexist city are discussed in Chapter 3. One area in which little systematic geographical research appears to have been carried out concerns the origins and implications of differences between male and female 'activity spaces'. For instance, it seems that little girls are allowed far less freedom of movement than little boys, in part because of the apparently higher chance of sexual attack they face. In adolescence, the identification of public space with boys, and private space with girls, seems to become even more marked. Street gangs, for example, are usually exclusively male. The effects of this on girls' and boys' knowledge of, and attitude to, the outside environment is likely to be considerable and needs careful study. Strong pressures are exerted on women to restrict themselves to the domestic areas of cities and urban life. These range from ways of restricting their mobility (from constraining clothes and high-heeled shoes, to jokes about women drivers) to an ideology that encourages women to consider themselves physically frail. This, however, is not to deny the real problems that arise from vulnerability to rape, which influence how women use space. The topic of the extent of women's fear in different types of environment, at different times and with different people, is an under-researched area of geographical research.

Because patriarchal attitudes and practices so clearly restrict the nature of the activities that women can undertake, and also the places where they can carry them out, radical feminists are united in seeing men as the enemy in the struggle for women's liberation. For this reason they are extremely suspicious of efforts to link the struggles of feminists with broader political movements, remembering only too well the way in which women and women's demands have been treated by men in movements such as those for civil rights and socialism. Radical feminists are also far more likely to argue for complete separatism from men, at least when it comes to organizing and participating in their own political activities and campaigns, so that women build up the necessary self-confidence to do such things.

Socialist feminism

In Britain, many women who worked actively in left-wing politics wanted to respond to the challenge thrown up by radical feminists in the late 1960s and early 1970s while maintaining some form of association with left-wing parties. As Coote and Campbell (1982, p. 31) put it: 'Socialist feminists saw their feminism as the human face of the left . . . (as) a critique of the male chauvinism of the left, which would transform much of its conduct and many of its priorities.'

Socialist feminists also moved away from a view of history, and geography, in which patriarchy is portrayed as some kind of constant which does not vary according to other characteristics of the societies in question. In particular, socialist feminists in Britain wish to understand the relationship between the struggle among different classes over who controls the means of production (the raw materials, machines and tools used in the manufacture of commodities) in a given society, and the specific role which women play in that society. They reject the notion that 'men are the enemy', arguing that women and men should unite in class struggle. However, this does not mean that women should accept the 'politics of deferment', waiting for any specific moves towards their own liberation until 'after the revolution'.

There have been a number of attempts to bring together two perspectives on women's position, first as subject to male domination, and second as participants, with men, in the class struggle. Whereas, in feminist theory, the family is accorded an important role in the maintenance of patriarchal relations, in marxist theory it is considered primarily in terms of its role in the *reproduction of the labour force*. To understand the concept of reproduction of the labour force, it is first neccesary to briefly outline some basic marxist arguments.

Capitalists own and control factories and other facilities for producing goods, and need a pool of workers on which they can draw if the factories are to operate. These workers sell their labour and skills to the capitalist for a wage, and their labour is then applied to the production of various commodities which the capitalist then sells. However, the value of the commodities made is *greater* than that placed on the workers' labour by the capitalist, and hence paid to them in wages. This ensures a surplus, which accrues to the capitalist once the goods are sold. The surplus is normally realized in terms of a monetary profit, some of which is used to reinvest in production facilities so as to make a further profit – that is, so as to accumulate capital – some of which is used, via taxation, to pay for the many public services and facilities that we expect the state to provide in advanced capitalist societies. In marxist theory, therefore, the labour of workers is the ultimate source of value (however it is measured) in capitalist goods and commodities. Thus for this process of production to continue workers must be clothed, fed, sheltered and rested while not at work to enable them to be fit to return to work each day, and some provision must also be made for the longer-term needs of the capitalist as workers eventually grow too old to work, and die. This last need is met, at least in part, by the fact that workers have children who grow up to become workers in their turn. Reproduction of the labour force, therefore, involves all the activities of caring for workers and of the upbringing of potential workers. If we consider the role of the family in this light, it is clearly a vital element in the reproduction of the labour force. Furthermore, if we accept the widely-documented fact that women are primarily responsible for child care and domestic work in all capitalist societies (and in virtually all other existing societies as well), then women in their domestic role have an important, although unpaid, place in the maintenance of capitalism. Put at its crudest, the men go out to work and are exploited there by the capitalist, and then return to their homes and families where they in turn subordinate and oppress 'their' women.

Capitalists thus benefit not only from exploiting the labour of waged workers, but also indirectly from the cheap labour provided by women in reproducing the labour force. It is assumed that such labour is paid for out of men's wages. Men, although exploited, benefit from the fact that women are primarily responsible for domestic work and the family. As already described, not only are the isolation, and many of the stresses and strains involved in this domestic work, largely avoided by men, but women find themselves in a position where they, together with their children, are legally and economically dependent upon men.

However, many women not only play a vital part in the reproduction of the labour force, but they also do waged work – often in the home. As some socialist feminists point out, in *theory* the woman does not need to work because she is supported by her husband's wage, but in *practice* vast numbers have to do some kind of paid work in order to make ends meet. As discussed in Chapter 4, the number of women in paid employment has risen sharply in Britain, and in other advanced capitalist societies, since 1945. The classical marxist view, as expressed by Engels (1972), is that women, like men, can only move towards full legal and economic emancipation when they become part of the labour force and so can participate directly in class struggle. However, this view takes no account of the various ways in which *men* benefit from the subordination of women. It simply sees men as agents of capitalism (e.g. Marshall, 1982) who, by confining women to unpaid domestic duties, reduce the costs of reproducing the labour force *and* provide a cheap source of labour, either to carry out jobs that men are unwilling to do cheaply, or to move into employment when there is a shortage of male labour, or both. It is certainly true that capitalists benefit from women's subordination in the way that classical marxist theory suggests. However, we emphasize that it is also in the *immediate* interests of men to avoid the major responsibility for domestic chores and child-rearing, to ensure that they get the better paid jobs, and that this factor must also be considered in analysing women's employment patterns.

During periods of economic growth – in particular the period before the current economic recession got fully underway – the segregation of employment by gender reserved higher paid jobs for men, while women's waged work provided valuable and often essential supplementary household income. During the recession, some women have become the major, or even the sole, wage earner in many households. In these cases, in terms of 'capitalist logic', men have 'priced themselves out of the labour market', and the previous alliance between patriarchy and capitalism to ensure a 'family wage' for men has been shown to be only in the immediate, rather than the fundamental, interests of male workers. This is not to argue that women simply benefit under these circumstances, since any reduction in household income directly affects them too, and there is no guarantee that the 'role reversal' which is much discussed in the press actually occurs on more than a very limited scale. Also, we must stress that women, and other subordinate groups like non-whites, earn lower wages and are more liable to be thrown out of work during a recession. The pattern, however, is very complex, varying by sector of economic activity and by region. The importance of understanding the role of female labour in the regional analysis of employment trends is illustrated in Chapter 4, which discusses industrial location and regional change. It shows that not only have changes in the organization and location of industry generated a high demand for female labour in Britain since 1945, but the characteristics (age, marital status, location etc.) of potential female workers have in turn shaped these changes. There is enormous scope for geographical research here, both on the lines illustrated in Chapter 4, and in terms of investigating the way in which the detailed fabric of women's and men's lives is related to such variations.

Both radical and socialist feminists have recognized the important role played by the *state*. The women's movement has formulated seven specific demands, all of which could require state action. They are: equal education and employment opportunity; 24-hour child care provision; free contraception and abortion on demand; legal and financial independence; freedom to define our own sexuality; freedom from rape and male violence against women. Potentially these demands would, if implemented, have far-reaching implications, not only for the relationships between women and men, but for the entire social and economic order. This is one reason why many in the women's movement have looked to the state as the appropriate instrument for social reform. Another reason is that the state, whether in the form of central government and various public bodies like nationalized industries and services, or in the form of 'the local state', for example local authorities, is a major employer of female labour. Women are employed as clerical assistants, as nurses, teachers, and in a range of manual occupations such as cleaning and cooking. However, we will defer consideration of the role of the state until we discuss the links between feminist and marxist approaches in geography.

Because of the way in which society currently operates, feminism can be regarded as an example of a *critical* current in social science thinking. Such approaches are characterized by the belief that the way in which society works is not beyond the control of the people who make up that society. In particular, it is argued that within any society the potential for change exists, and that ordinary people must get together to ensure that such changes are under their conscious control. We shall now briefly examine two sets of approaches in geography which, like feminism, have been critical both of the relationships of dominance and subordination that exist in society at large, and of the way in which geographers deal with such relationships. The similarities and differences of these approaches to feminism will be considered.

Marxist approaches

The essence of marxist approaches is that capitalist societies, such as that in Britain, are characterized by in-built contradictions which means that antag-onism along class lines is bound to occur. This is because, as we outlined earlier in our discussion of capitalism in the section on socialist feminism, a numerically small class (capitalists) exploit the labour of the many (workers). However, marxists in geography have been chiefly concerned with the production process itself, with alternative approaches to economic location theory, and with the production and consumption of publicly provided services like housing and education. As Bowlby *et al.* (1982, p. 20) point out, marxist geographers have had little to say about the reproduction of the labour force or the oppressions of everyday life – whether affecting women or any other group. Although marxists and other socialists in geography have often sympathized with, and at times supported, feminism *outside* their academic work, there has been little evidence of it *within* their academic work. The only exceptions are socialist feminists, mostly women, and most of whom in Britain play an active role in the IBG Women and Geography Study Group.

There seem to be two areas in which marxist approaches in geography offer most help to feminist analysis. The first area in fact represents a blend of *materialism* and the notion that ways of living are *specific to a particular time and place*. Materialism refers to the notion that society is impossible to understand, or change, unless we analyse the social and technical modes of organization which are involved in catering for basic human needs (like food and shelter), ensuring biological reproduction of the species, and providing the generally accepted 'standard of living' that prevails in that society. Marxist geographers also stress that the way in which material needs are catered for can vary enormously over time and space. Thus, the way in which people maintain a particular standard of living in capitalist societies depends crucially upon the relationship between capitalists and workers. As we have seen, the degree of control that women can exercise over their sexuality and over their own bodies, as well as over opportunities to earn a living, is also materially limited by patriarchal relations. However, capitalist–labourer and male–female relationships are not the same even within the same capitalist economy (see Chapter 4) and may differ considerably from one capitalist country to another. Why, for example, should child benefit and child care facilities be provided on such a generous scale in France, and not in Britain, when both are advanced capitalist states? If we consider differences between countries at various stages of development within the capitalist world, these are even more striking, as Chapter 6 clearly illustrates. The same chapter also gives some idea of the differences between capitalist and non-capitalist countries in the developing world. Similarly, Chapter 3 not only emphasizes the spatial segregation of men's and women's work that arose with the development of the industrial city, but the various forms it took over time and space, for different strata of society, as capitalism changed.

The other area in which marxist geographers have recently shown strong interest is the study of the *state*. Conceptions of the state vary from that of an 'apparatus' which is the instrument for furthering the interests of the dominant economic group in society – for example, the capitalist class in capitalist societies – to that of an essentially neutral institution which encapsulates and responds to the various conflicts and pressure groups in society. At one end of the spectrum, the instrumental conception of the state implies that for radical change to occur, the complete destruction, or, at the very least the take-over, of the state apparatus by a different class (in classical marxism, the workers or proletariat) must be achieved. At the other end of the spectrum, the view that the state is relatively neutral and autonomous implies that changes can be achieved within the current framework of the state. Thus some groups, such as many feminist groups in Britain during the 1970s, organize, agitate and pressurize for reform via legislative change. Socialist feminists have asked for, and in some measure obtained, support in these attempts from various left-wing parties and the Labour Party, and from trade unions. As the introductory chapter has pointed out, some progress in terms of government legislation was made during the 1970s, but these gains have begun to be eroded as the current recession has deepened. Profits are reduced and 'society' is less willing to accede to demands for such things as equal pay, free or subsidized child care and better medical facilities. Wage squeezes and public expenditure cuts by the state have been strongly in evidence in Britain during the last three or four years. Socialist feminists, in particular, see a clear connection between class struggle – in which workers resist cuts in their living standards which are imposed directly by capitalists, or indirectly by the state – and the battle for women's liberation. Feminists, however, take part in this struggle only on condition that their demands are recognized as a central part of these campaigns. Unless this condition is met, women's subordination simply becomes another aspect of, and is reduced to, class struggle. The important role played by men in the subordination of women as expressed in the concept of patriarchy, is diluted or lost.

Like marxists, feminists wish to see fundamental changes in society, particularly as these would affect the position of women. There are, however, considerable differences among feminists in the scope of the changes they envisage, and in what they consider the implications of such changes to be in terms of the existing power structure in society. Some feminists are basically reformers, wishing to improve the position of women while leaving the structure of capitalist society essentially unaltered. Others argue that no really significant improvement in women's position can occur without the destruction of capitalism. Some suggest that the implications of the reformers' demands are, in fact, revolutionary, even though those feminists may think of them in terms of constitutional reforms. We cannot answer the questions raised by these alternative views here, but we hope you will discuss them as they are questions which frequently underlie discussions about feminism.

Finally, it is very important to realize that feminists have criticized the organization of parties, and particularly left-wing revolutionary parties. They see such parties as hierarchical and undemocratic. A clear statement of these criticisms can be found in *Beyond the Fragments* by Rowbotham *et al.*, (1979). It is argued that parties and party leaders act as an elite, drawing, in a dogmatic fashion, upon the analysis provided by 'scientific socialism' in order to control people; not doing things *with* them, but *to* or *for* them. This is held to represent a way of doing things which is typically 'male'. Sayer (1983) argues, however, that a form of humanistic marxism exists which is compatible with the kind of socialist feminist practice defended in *Beyond the Fragments*. This approach recognizes that people have to be organized if they are to struggle effectively for their rights, but considers how this might be achieved without the distortion and repression of human relationships by power. Consequently, particular attention is paid to forms of organization which do not perpetuate the way in which men currently oppress women through patriarchy. Furthermore, the traditional emphasis in marxism on battles between capitalists and workers over the production process is modified, to include struggle to change the entire way of life of people living in capitalist societies; so allowing explicit attention to be given not only to the public position of women and other oppressed groups, but also to the 'private sphere' as an area which is intrinsically political. As many feminists have put it, 'the personal is political'.

Phenomenological and humanistic approaches

These approaches to geographical study have attracted considerable interest since the early 1970s. Both are characterized by dissatisfaction with mainstream 'scientific' geography which, they argue, reduces the study of human behaviour to that of a subject (the scientist or researcher) contemplating an object (a particular pattern of human spatial behaviour). It is argued that most conventional methods of data collection in social science impose the observer's framework of thought upon the 'objects' of study, rather than revealing what they really think. A similar point is also made by feminist writers such as Anne Oakley (1982). Like feminism, these phenomenological and humanistic approaches emphasize the importance of caring and nurturing in relationships with other people and with the environment, and also the crucial role of day-to-day feelings and experiences. These may not only be taken for granted and ignored by researchers, planners, bureaucrats and others in big business and big government, but also by the individuals who are themselves the subjects of such feelings and experiences. The kinds of approaches advocated by phenomenologists and humanists – such as in-depth interviewing and consciousness-raising sessions, use of the insights afforded by novels and other literary sources, and an emphasis on 'soft' or qualitative data – can, and have been, developed by feminist researchers (see, for example, Roberts, 1981, and Chapter 7 in this book). It is worth pointing out here that one very influential feminist study, *The Feminine Mystique* (Friedan, 1968), brilliantly captures the mood of many white, middle-class American women, living without significant material deprivation but living through and for their husbands and children rather than for themselves, by making considerable use of informal interviews and anecdotal 'experiential' evidence. Feminists have also used literature as a vehicle for exposing and developing new ideas. For example the experiences and feelings of women in suburbia, again in America, have been conveyed by Marilyn French in her novel *The Women's Room*. Marge Piercy's novel *Vida* shows how a woman in a left-wing revolutionary group unknowingly acquiesces in being treated as a 'second class citizen' by her sexist male comrades, and the experiences of a black woman from Mississippi who comes into contact with black and white civil rights activists in 1960s America are the subject of *Meridian* by Alice Walker (1982).

Although humanists emphasize that the 'ordinary lives of ordinary people' should not be ignored, feminists argue the case with far greater passion. They stress the way in which ordinary women are oppressed and produce accounts of, and explanations for, the roles women have played, and do play, with the aim of ensuring that women are no longer 'hidden from history' or, for that matter, from geography. It is also striking that, although phenomenologists and humanists emphasize caring, humanity and gentleness in person-to-person and people–environment relationships (and these attributes, together with the reaction against 'hard-nosed' scientific and rational approaches are regarded as traditionally female virtues), they nearly always talk of 'men' and pay virtually no attention to the role of women, let alone the women's movement (e.g. Ley, 1983; Seamon, 1979). Buttimer does have something to say on these issues (Buttimer, 1972; 1976a), and has also produced a paper which explicitly criticized 'sexist rhetoric' (Buttimer, 1976b). However, the contributions from phenomenologists and humanists on women's place have been meagre.

Approaches advocated by phenomenologists and humanists can be used to examine the way in which the carrying out of particular activities, and the associated use of particular places and spaces, come to be regarded as 'feminine' or 'masculine'. For example, 'women's space' is inside the home, waiting for children to come out of primary school, or at the shops, while 'men's space' is at work, in the pub or digging the garden. This division of labour and space is not totally rigid, of course, but sufficiently marked that certain activities and spaces come to symbolize femininity, others masculinity. An individual feels that frequenting particular places and undertaking particular activities constitute what it means to be female or male. Such feelings may run counter to actual behaviour – for example, most women do some kind of waged work, but tend to feel that going out to work is not an integral part of being a woman, whereas it is for men. We shall point out here that many studies of everyday behaviour in the social sciences are concerned with males rather than females, particularly those that focus on youth cultures and subcultures (McRobbie, 1980), demonstrating that the neglect of women is by no means confined to geography.

The strength of phenomenological and humanist approaches, and of certain strands in feminism, is at the same time a potential source of weakness. Unlike most marxist thinking in geography, there is a strong emphasis on the importance of understanding the everyday lives and thought of ordinary people. Furthermore, it is argued that socialists, capitalists, experts and planners constantly interfere with ordinary people who simply wish to get on with their own lives; and who, if left to themselves, would make a far better job of it (eg. Relph, 1976; 1981). Thus, if people were allowed to stop and think about the meaning of their everyday lives, perhaps in free-ranging discussion with others in a non-hierarchical setting (no group leaders, no chairpersons etc.), then it would become possible to build on this developing self-awareness so that people could help themselves through various community movements. However, a major drawback with such approaches is that there is a tendency so to glorify ordinary, everyday experience, that there is a failure to move beyond the necessary stage of description, self-revelation and empathy to an explanation and analysis of the reasons for those experiences; this in turn is related to a failure to consider how power relations and inequalities in society are interlinked. In fact, it could be argued that phenomenologists and humanists are basically modest reformers, looking for ways in which the ordinary person may better their lot within the chinks and interstices left between the great power relationships in society, adjusting to the prevailing patterns of dominance and subordination. We should also add here that phenomenologists and humanists tend to show a *general* concern for the way in which ordinary people are subject to various forms of authority, rather than analysing the specific forms of exploitation and oppression that occur. Their insights are, therefore, of limited value to those interested both in analysing the specific forms of oppression to which women are subjected, and in using this as a basis for removing such oppression.

Theory and practice

One of the central tenets of critical approaches to social science is that we should not only properly understand society (theory), but use such understanding as a basis for a programme of action (practice) to change society. If we succeed in changing the position of women, this requires a reassessment of the situation (theory), and so on. This change is only possible through an endless cycle of theory and practice. As we have seen, theory and practice relate to the 'nitty-gritty' problems of how individuals live their lives, how feminist groups organize, and to the overall social, economic and political structuring of society. Modern geographical work ranges over all these aspects. In the following sections we analyse the relative locations of home and work in the city, the day-to-day work of women in the labour force in terms of the requirements of capitalists, access to health care in relation to reorganization and centralization of health services by the state, and the differential impact of 'development' on men and women in Third World countries. These topics have been chosen as *examples* of the way in which a feminist approach to topics traditionally seen as part of mainstream geography changes and deepens our analysis. We hope that these examples will suggest other areas in which feminism and geography have something to offer one another.

Further reading

1 On the relationships between feminism and marxism, see L. Sargent (ed.), *Women and Revolution* (Pluto, London, 1981), particularly the introduction and the paper by H. Hartmann, pp. 1–41; and R. A. Sayer, 'Beyond the Fragments and the meaning of socialism,' 'Meeting of Minds', in Proceedings of a one-day conference (IBG Women and Geography Study Group, 1983).

2 For a general discussion of the political, legislative and economic context of the women's movement in Britain, see A. Coote and B. Campbell, *Sweet Freedom* (Picador, London, 1982).

Topics for discussion

1 Discuss the approaches of socialist and radical feminists and consider the implications of these approaches for geographical study. What are the strengths and weaknesses of each approach? Which approach do you think is correct?

2 What similarities and differences are there between feminist geography and other critical approaches in geography? How important are the differences?

3 Keep a record of all the places that you visit, and all the activities you undertake, for half a day. Did you visit any places or carry out any activities simply by virtue of your gender? Your sex? Would you have done any of the activities differently, or simply done different things, if you happened to be of a different gender? How would you explain any differences?

PART II

Four examples of feminist analyses in geography

3 Urban spatial structure

Introduction

During the last 150 years there have been major changes in the land use structure of most of the world's cities, in the location and organization of paid work, in women's participation in the labour force and in their domestic work. The revolutionary transformation of land use that took place in industrializing western cities is a familiar theme of urban geographers. The rapid expansion of urban population, the growth of suburbs with distinctive class characteristics, the improvement of transport, the development of specialized shopping and service centres are part of a well-known story and form the subject matter of most urban geography courses. Although one of the most obvious features of these changes has been the growing separation of home life from paid employment geographers have, in general, paid only cursory attention to the implications of this division. They have neglected to analyse both the impact of these changes on social relations between men and women, and the relationship between the organization of domestic work and urban spatial structure. Yet, as we will explain later, changes in domestic work and in the functions of the family have been an essential part of the spatial development of the modern city.

The process of industrialization ensured that domestic work carried out by women became less and less concerned with the making of goods for home consumption, and more and more concerned with using goods manufactured elsewhere to *service* other household members. For example, a typical North American married woman in 1800 would have lived on a farm, made her family's clothes, soap and candles, baked her own bread, looked after the kitchen garden and raised animals and poultry for food. By 1950 she would have lived in a city, bought manufactured clothes and cleaning materials, used electric light and bought canned, packaged and frozen food. Furthermore, while much of the rural woman's work would have been done in the company of husband and children who were also working on the farm 150 years later her urban counterpart would have worked largely in isolation while her children were at school and her husband earned wages outside the home in a downtown factory or office. Thus the Industrial Revolution involved a transformation of the family from a joint productive unit, in which future workers, the children, were trained largely by their parents, to a reproductive unit devoted to caring for men or children whose work, or training for work, took place outside the home in spatially quite distinct and separate areas. This transformation of domestic labour was part of a process of change in *all* work, and we will argue later that any proper understanding of the growth of our present capitalist society must include an understanding of the role of domestic work.

We do not present life in the pre-industrial rural family as ideal. Work was hard and wives did not have a position of equality with their husbands; 'men's' work was distinguished from 'women's' work and considered of higher status. Nevertheless, women's labour was seen to be difficult and demanding and was recognized as 'real work', vital to the survival of the family enterprise – a recognition that was no longer accorded to the domestic work of the woman of the 1950s. Today in the advanced industrial societies of the western world, people who 'work' go *out* to work, whereas the home is normally portrayed as a place of rest and respite from labour. Domestic work and home activities in general, are considered by most people, including academics, as dependent on, and of secondary importance to, activities outside the home. Yet, for most of us, the family and the home are still central to our lives.

The vast majority of people live in nuclear families in single-family housing. Only a small minority live in non-family groups, in hostels, hospitals, or other institutions, or are homeless. A large proportion of everyday life is spent at home; admittedly much of it asleep, but about half our waking activities are carried out within or close to the home. For young children, the elderly, the unemployed and for housebound women, the proportion of time spent in the home is much higher. For all women, but particularly those with small children, their home is the location of both work and recreation, they eat, sleep and 'rest' at the scene of their labour, and for the most part, this labour is entirely solitary.

Despite the significance of home-based 'private' activities in people's lives, particularly the lives of women, most human geographers would include in a list of their subject matter only those activities which take place *outside* the home, in the 'public sphere'. Geographers and planners, while unable to ignore the spatial separation of waged work from home, continue to take it for granted. Planners zone different areas of towns and cities for residential development and for employment which, as we shall argue later in this chapter, results in the isolation of women who do not participate in the urban labour market, and raises problems of timing and organization for those who combine waged with domestic labour. Academic geographers have also failed to question the gender implications of the separation of home and work.

In the next section we briefly examine the major explanations of urban spatial structure with which you are, or will become, familiar from courses with such titles as 'residential location theory', 'urban geography' or 'the geography of housing' and such recent books as those by Bourne (1982), Bassett and Short (1980) and Dicken and Lloyd (1982).

Urban theory

The models of the Chicago school (Burgess's rings, 1925, and Hoyt's sectors, 1939) and those of neo-classical economists (such as Alonso, 1960, and Muth, 1969) are based on the idea of competition between different types of activities and between individuals with unequal resources. The 'dominant' activities (in Chicago terminology) or those valuing centrality over space (in the economists' models) – office employment and certain types of retailing – occupy the city centre, and the 'less competitive' – such as housing – the city outskirts. Housing itself is also spatially differentiated – households of different social class and income group live in different areas of the city. More recently-developed work on urban spatial structure, such as that based on the idea of urban gatekeepers (Rex and Moore, 1967) and on marxist ideas of the relations of production in capitalist economies (Castells, 1977; Harvey, 1975, 1978) has 'unpacked' the notions of competition and dominance. It has shown how the actions of the local and central state and of the industrial and financial institutions in modern societies structure the operation of the land and housing markets.

What unites all these approaches is their neglect of the implications of the spatial separation of men's and women's work. Domestic activities are ignored. Indeed, until recently, women's activities have not even been distinguished explicitly from those of men. In traditional residential location theory, the unit of analysis is the household which typically is assumed to be a male-headed family unit. The economic models, based on a trade-off of suburban space and accessibility to centrally-located employment opportunities, assume only one waged worker – the male head – and ignore the additional trade-offs that take place in dual-career families and indeed between the location of work, schools, shops and other important centres for different household members. Thus the household is seen merely as a statistical unit, without an internal hierarchical structure of power relations.

In ways which we shall describe below, feminist geographers are currently attempting to understand the basis for, and consequences of, the separation of 'male' and 'female' spheres in cities, the division of 'public' activities from 'private' activities, and to place an analysis of the changing relationship between domestic and waged labour at the centre of a feminist urban theory. From this work it is becoming clear that the present organization of urban land uses tends to benefit employed men with wives, at the expense of both married and single women struggling to run homes, care for children and, in greater and greater numbers, also manage a part-time or full-time job. We believe that the changing experiences of urban women in different social classes, and their reactions to these changes in struggles to change their lives, should form part of the subject-matter of urban geography.

Existing feminist work on urban issues can be grouped into four broad categories. The first group comprises studies that re-examine the development of cities during the industrial revolution from a feminist perspective. We shall be drawing on this work in the next part of this section and references to it can be found in the reference list at the end of the book and in Further reading on page 65. The second group is concerned with analysing the situation of women in cities today. One popular topic is women's access to urban goods and services – something we look at later in this book (see Chapter 5). Examples of other topics are: women's position in the housing market; domestic work and housing design; women's activity patterns in time and space; women's and men's different perceptions of the urban environment. Examples of North American work in this vein can be found in a special issue of the journal *Signs* published in 1980 (vol. 5, no. 3), and in a book published in 1981, *New Space for Women*, edited by Werkerle, Peterson and Morley. This book also contains examples of studies which fall into our third grouping which covers work on women's impact on urban design. Here we find studies of individual women as designers and architects, and also of the part that is or could be played by women in general in the professions of planning and architecture. Not surprisingly, much of this work emphasizes women's lack of representation in the urban design professions. Recent British ideas in this area can be seen in *Making Space: women in the man-made environment*, published by Pluto Press (1984), and edited by the Matrix Book Group, a group of feminist architects and planners. Finally, there are speculations about what a non-sexist city might look like – a theme which we will examine briefly at the end of this chapter.

Now we turn to look at the relation between changing domestic work and the ways in which the internal spatial structure of industrial cities developed during the nineteenth and early twentieth centuries.

The separation of home and 'work' in the industrial city

Urban geographers have recognized three distinct phases leading up to the modern city of the twentieth century. Paul Knox in a recent book (Knox, 1982) terms them 'The Pre-Capitalist City', 'The Transitional City' and 'The Victorian City'. In this section we will outline the main features of the interactions between changes in the form and organization of the city, and changes in the nature and location of domestic and waged work in each of these three phases of urban development. We assume some familiarity with conventional urban historical geography, to which Knox's chapter on urban development forms a good introduction. We will be drawing on work on Toronto by Mackenzie (1980), on London by Rose (1981), on Birmingham (England) by Hall (1982) and on Philadelphia by Miller (1983). Since the start and rate of industrialization differed between England, America and Canada and also between different regions in each country, no precise dates for each phase are given. However, perhaps it will help to explain that we start our story towards the *end* of the pre-capitalist phase just before cities were transformed by the impact of industrialization. This places us roughly in the mid to late eighteenth century in Britain and the very early nineteenth century in North America.

The pre-capitalist city

The pre-capitalist city was tiny by today's standards – a 'walking city' without any purely residential, commercial or industrial areas. It housed four main groups: the *land-owning aristocracy*; the *merchants*, whose trading activities provided work for such people as clerks and lawyers; *skilled artisans* producing goods largely for local consumption; and *unskilled workers*. The latter were only a relatively small proportion of the total – for example, in pre-capitalist Toronto they formed less than one-fifth of the population. Most districts comprised a jumble of land uses housing people from all walks of life, although higher status people tended to live near the centre.

For the artisans, home either *was* the workshop or was physically close to it. Wives often helped in the business and even where they did not, they lived surrounded by the same activities that were part of the daily experience of their husbands. Male children were educated at home and then through apprenticeships in the family of a skilled master; female children were taught household work at home. It is worth emphasizing that the term 'family' at this period was often used interchangeably with the word 'household' to include apprentices and other workers who lived under the same roof (Barrett and McIntosh, 1982). Hall (1982) gives us a glimpse of this way of life, just as it was ending, in the shopkeeping family of Richard Tapper Cadbury who lived in the centre of Birmingham in 1800. Shopkeepers and artisans in the pre-capitalist city lived in a very similar style – indeed many workshops were also retail shops. Elizabeth Cadbury, Richard's wife, helped in the shop, looked after affairs when her husband was away buying new stocks, and organized a household of apprentices, female shop assistants, children (she had ten – eight of whom survived), her elderly mother and two female servants.

The way of life of the skilled artisans, small merchants and traders bore some similarity to the pre-industrial rural family. Both were soon to vanish almost entirely. It was among the families of the poor and the rich that the outlines of the family relationships of the future were to be found.

The families of the poorest unskilled labourers generally lived outside the city centre. Poverty obliged all who could, including children, to go out and work for wages. Married women found jobs as seamstresses or maids of all work while their husbands did unskilled labouring work. Mackenzie terms this type of family a 'unit of survival', but although its members' lives depended on one another it was in a very different fashion to the interdependence of either the rural or the artisan family – for they were not working in their own productive enterprise, although many grew vegetables and raized poultry primarily for family consumption. In the main, however, they worked separately and independently outside the home.

Among the families of the aristocracy and rich merchants a very different pattern of family life was emerging, involving not only the separation of home and workplace but also the separation of women and children from involvement in 'work'. Women in such families had always been important in bringing in property and money through their marriage settlements and kinship ties. By bearing heirs they ensured that family wealth would be preserved. Inside the home they managed substantial households and helped to educate their children, while outside the home they had an important role to play in philanthropic and charitable activities. The view that women were unsuited to any work other than domestic, had begun to gain ground among the upper classes in England by the seventeenth century. Thus, although they might contribute to the family wealth through marriage, such women were rarely directly involved in its creation. Moreover, as trade grew in importance, men increasingly began to carry out their business outside the home and these women's everyday links with the world of business became more tenuous.

This pattern of the non-working dependent wife and children was, as the way of life of the rich, soon seen as an index of wealth and prestige and adopted by families of the growing ranks of small merchants, professionals and clerks as well as by the wealthier artisan families.

The transitional city

New transport networks, canals and railways, increased the integration of the economy and stimulated trade and production in both craft workshops and new machine-based factories. As yet the new technology had not undercut the majority of skilled craft workers and the new factories were still small. Large numbers of rural in-migrants came to work in the expanding manufacturing and construction sectors. The numbers of unskilled labourers grew until they made up nearly half the population. The old landowning aristocracy became less important economically and socially, and their place was taken by the factory owners and wealthy professionals whose activities now dominated urban economic life.

The first specialized commercial and high status residential areas were established. In Toronto, for example, two new high status, purely residential areas grew up equipped with prestigious churches, schools, doctors' surgeries and colleges which complemented the family's role in educating and socializing the future elite. Although city cores still remained areas of high residential status they were becoming increasingly dominated by commercial and industrial land use. In Birmingham, Richard Cadbury's son John, who had set up a tea and coffee shop next to his father's drapery, also started a small cocoa factory round the corner. At first, like his parents, he and his wife lived over the shop, but in the 1830s they moved to the new suburb of Edgbaston whose middle-class residential character was assured by restrictive leases preventing the building of workshops in gardens or the opening of shops on the premises.

The majority of the expanding unskilled labouring population was still generally found on the outskirts of the city although their work was predominately in the centre. This often required relatively long journeys to work which, at this time, still had to be made on foot. Feminist historians have already done a substantial amount of analysis of the changing patterns of women's employment in Europe in the nineteenth century and its importance to the working-class family (Scott and Tilly, 1975). This shows that women's wages were generally less than those of men (in Toronto they were only one-third to one-half those of adult men) and they worked in the least attractive jobs – in laundries, rag mills and sweat-shops. An important feature of married women's employment was that those who worked were despised as morally inferior by their non-working middle-class sisters. For by this time the pattern of dependent wife with her children, living a life centred on the home and devoting her spare time to 'good works', was not only thought of as an index of her husband's wealth, but also as *natural* and the ideal way of life for married women of *all* classes.

Women were now held to be morally superior beings, innately gentle, loving, asexual and tender but also over-sensitive, irrational, emotional, delicate and weak. Their proper role was in serving men either as wives or servants. The home, wife and children were seen as a refuge for men from the brutalizing influence of work. Home and female society constituted the private sphere in which emotions could be expressed; work and male society constituted the public sphere of rationality, science and the intellect. These two spheres were becoming, both literally and metaphorically, separate areas of life. Davidoff *et al.* (1976) have shown how this ideal of the home as a haven of moral stability and order was also linked to the myth of the ordered, hierarchical rural community. Both were enshrined in the new middle-class suburb in which the houses were built as miniature versions of the English country house, while within each home the household of husband, wife, children and servants was seen as a small scale version of larger aristocratic establishments and as an image of a stable and hierarchical social order.

As manual domestic work was regarded increasingly as socially demeaning and inappropriate for the middle-class married woman, so the desired way of life of the middle-class family became more dependent on domestic servants. During the second half of the nineteenth century domestic service became the most important single occupation for women in Britain and North America. It was particularly important for single working-class women who found independent survival difficult on the low wages paid to female industrial and agricultural workers. In England and Wales it accounted for over 40 per cent of employed women throughout the half century, while about 70 per cent of domestic servants were female. In America the comparative figures are 36 per cent and 66 per cent. In the British census of 1901 'domestic service' was the largest single occupational group. 'Going into service' was an important means of rural–urban migration, since the demand for servants was highest in the towns and cities where the new middle-class white-collar workers congregated (Figure 1), yet most geographical analyses of migration rates have neglected gender. There were substantial regional and interregional differences in the numbers and proportions of servants, which reflected both the degree of urbanization and the relative sizes of the middle-class and labouring populations (Figure 2).

Domestic service was thought by the middle class to be an ideal occupation for single women of the lower classes. Greg, writing in England (1862) about domestic servants, explained:

> they do not follow an obligatorily independent, and therefore, for their sex an unnatural career:- on the contrary. . .they fulfil both essentials of a woman's being: they are supported by and administer to men.
>
> (Quoted in Davidoff *et al.*, 1976, p. 168)

As we will see, the expansion of industry was to threaten the servant-based middle-class way of life. Furthermore, whatever the virtues of domestic service, it was not able to provide jobs for all women who needed them in order to survive. Throughout the nineteenth and early twentieth centuries both married and single women of the labouring classes were obliged to work in such 'unnatural' occupations as factory work, street selling and prostitution.

The Victorian city

As the factory system became the dominant form of manufacturing some time around the second quarter of the nineteenth century in Britain, and during its last quarter in North America, city populations exploded in size. The demise of the small workshop signalled the end of the old mixed residential and industrial districts. Home and work were now separate for the majority in all classes. As commercial and manufacturing activities expanded in the city centre, new, purely residential districts were established on its periphery. The commercial core was finally abandoned as a high status area and an area for family living. Cities expanded their areas, new suburbs were built (served by rail) and the first subsidiary shopping centres appeared.

A time–geographical study carried out by Miller (1983) on women's activities in the suburbs of Philadelphia during the mid to late nineteenth century suggests that less wealthy middle-class women living in the new suburbs, particularly those with large families, were cut off from city centre activities. Given the constraints of time and space, frequent social, shopping and personal business visits became difficult as did engagement in philanthropic activities. Their lives were necessarily more concentrated on the residential area of the suburb with its limited opportunities for non-domestic activities.

Lower down the social scale residential segregation was also increasing. In most cities many unskilled manual workers moved into multi-occupied housing around the core. However, some more affluent working-class families were also moving out to new suburban housing. In the years between 1890 and 1920, suburban living came to be seen as the 'ideal' family environment for *all* classes in both Britain and North America, and by 1920 a substantial portion of the working-class as well as the middle-class population in both countries was already living in suburbia. Two important factors in the process were contemporary worries about 'urban problems' and 'the woman question'. As we will see, these worries centred both on the family and the reproduction of the labour force.

'Urban problems'

Vast numbers of people who came to work for low wages in the new factories lived in appalling conditions. Reports of insanitary housing, neglected and delinquent children, drunken workers and epidemics of infectious diseases soon began to alarm the middle classes. Furthermore, the success of the factory system provoked unrest among the craft workers and socialist organizations spread through the working class. Thus the concern of the middle class was not only that infectious disease might threaten their own health or that slum conditions were inhumanitarian, but also that these conditions threatened the supply of adequately skilled and hard-working labourers. It was felt, and rightly so, that the working-class family was not able adequately to raise and educate its children for their future as industrial workers. While one solution to this was seen as extension of state-aided education, another was elimination

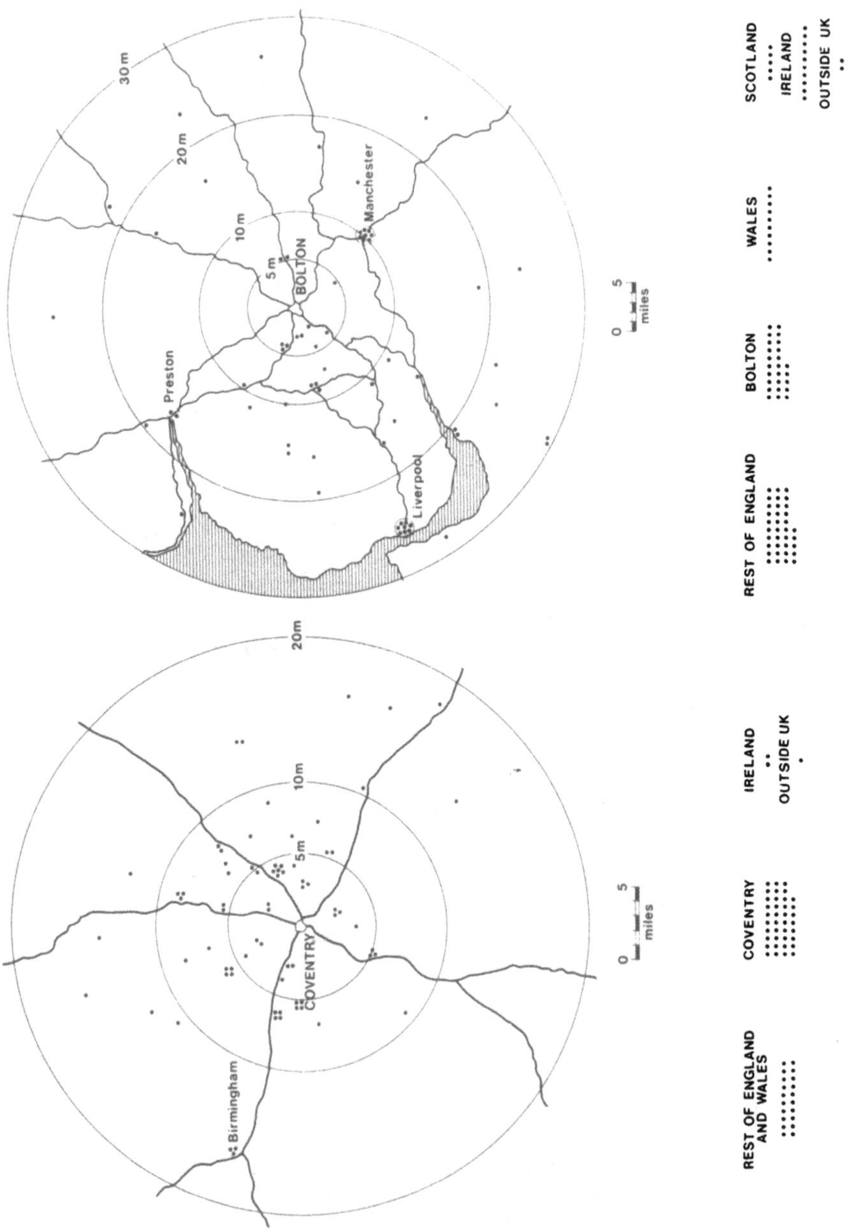
30m
20m
10m
5m
BOLTON
Manchester
Preston
Liverpool
0 5
miles
SCOTLAND
IRELAND
OUTSIDE UK
WALES
BOLTON
REST OF ENGLAND
20m
10m
5m
COVENTRY
Birmingham
0 5
miles
IRELAND
OUTSIDE UK
COVENTRY
REST OF ENGLAND AND WALES

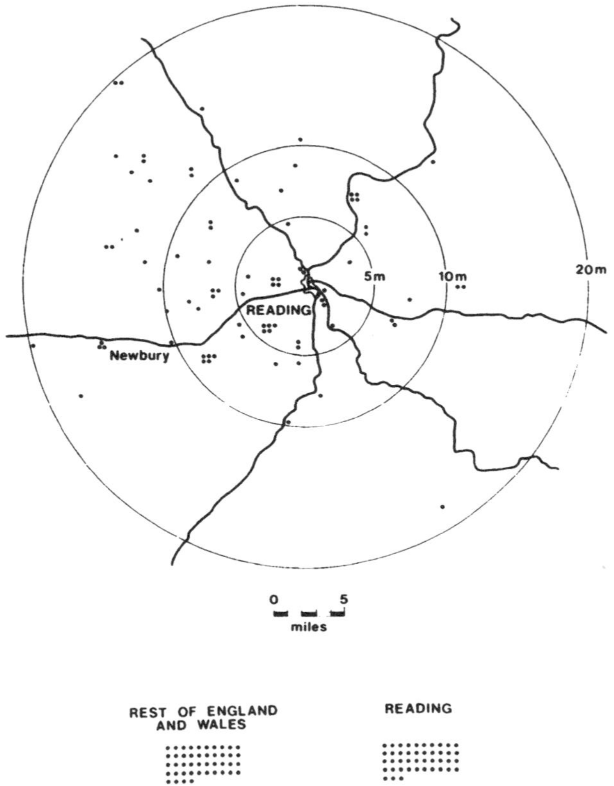

Figure 1 Birthplace of private servants resident in three British towns, 1871: contrasting patterns of migration to towns with different industrial structures and local hinterlands
Half the migrants to Bolton, a textile town in a textile area, came from other urban areas and over half came from well beyond the local hinterland. Four-fifths of the migrants to Coventry, a textile and watch manufacturing town, came from its immediate rural hinterland. Servants migrating to the distribution and food manufacturing centre of Reading were also predominantly of rural origin but a quarter came from beyond the local hinterland.

Source: Ebery and Preston, 1976, pp. 80–1

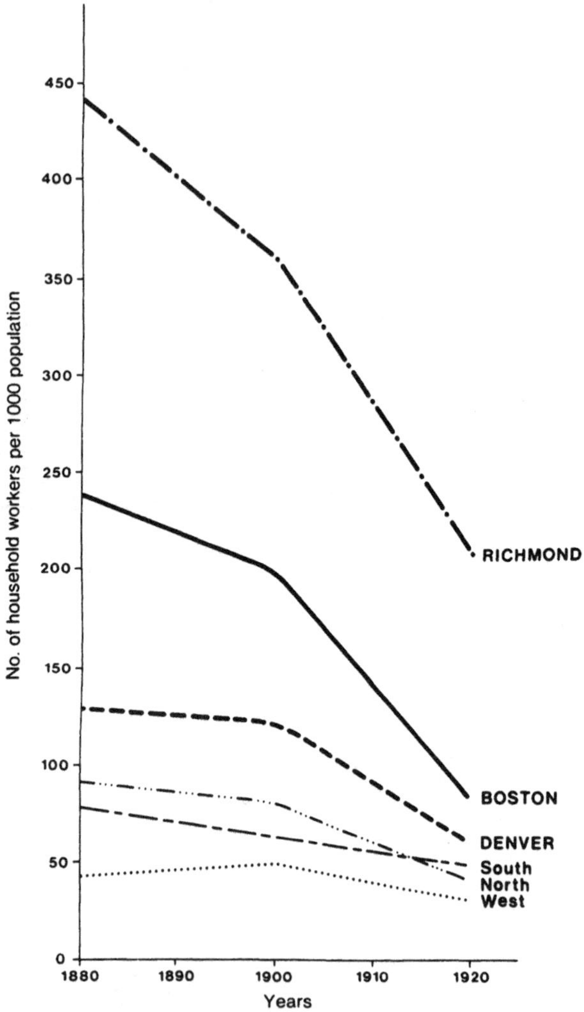

Figure 2 Contrasts between the proportion of domestic servants in different cities and regions in the United States of America over time.

Source: Hayden, 1981, p. 15

of slums and the provision of working-class suburban housing. For slum living was held to undermine family life and the inculcation of the virtues of thrift and hard work. 'The slums should be attacked and abolished because they are the greatest enemy of the home which is the foundation stone of the state', wrote J. J. Kelso in Toronto (quoted in Mackenzie (1980), p. 73).

'The woman question'

The home was also felt to be under attack from another direction – from 'emancipation' among women. Industrialization provided increased opportunities for women to earn their own living. Cheap female labour was attractive to employers, while women themselves preferred the relative freedom of office or factory employment to the rigours of domestic service. Many working-class women were employed in the manufacture of new household commodities (prepared foods, household machinery, ready-made clothing) while middle-class women could work in the expanding education and health sectors. In both cases not only were jobs for women being made available *outside* the home, they reduced the scope of (and women's control over) what went on *inside* it.

> 'Laundry interests', 'baking interests', 'canning interests', 'jam and preserving interests' have invaded the individual housewife's immemorial 'sphere', and have wrenched from her hands her exclusive control and responsibility for the health and well-being of the household.
>
> (Leathes, 1914, p. 74, quoted in Mackenzie (1980), p. 56)

Among middle-class women, dissatisfaction with their increasingly restricted domestic life was an important stimulus to campaigns for the vote and for access to higher education.

Women's new desire for emancipation, their movement into new occupations, the increasing cost of nursemaids and servants for the rich and the inadequate housing conditions of the poor were thought to account for the 'alarming' declines in marriage and fertility rates of the period. The 'ideal family' was under attack from new ideas challenging women's domestic role, from new household products and new 'female' paid occupations.

Opposition to this attack did not come only from the middle class. Working-class men also opposed women's work outside the home as 'unnatural' and protested against competition from cheap female labour. Their campaigns to improve wage rates and housing conditions were linked to demands for a 'family wage'. These were attempts both to improve the standard of living of all household members and to insulate 'the family' against the disciplines and uncertainties of wage working. But they were also attempts to create a 'dependent wife and children'.

Given prevailing ideas about the home and family and the view that poor housing 'caused' vice, delinquency and social unrest it is not surprising that the development of suburban housing for the working class was advocated as a solution by the majority of both middle- and working-class reformers. However, it is important to realize that there were people with differing views about the most desirable form of urban areas and we will discuss some of these at the end of this chapter.

Simply seeing suburbs as a desirable form of housing for the working class was not enough to bring about their development. Urban reformers, property developers and many businessmen pressed for some form of state subsidy to aid suburbanization. For example, in Toronto during the 1880s and 1890s, they succeeded in gaining municipal subsidies for the establishment of new transport lines, water supplies and sewerage and the extension of the city limits. It was argued that by encouraging private enterprise to provide suburban housing and cheap and rapid transport for the working class they would also provide: 'plentiful labour, well-housed, well-fed and intelligent labour. . .(which). . .means satisfactory conditions for the employer and manufacturer.' (McClean, 1909, quoted in Mackenzie (1980), p. 81). Suburbanization was thus thought to offer profits to private enterprise, and improvements in their labour force. However, it was explicitly recognized that the new suburbs were for the 'respectable' working class who would thus be separated from the influence of the dangerous and 'unrespectable' poor who would remain in the centre. Rose (1981) gives examples of similar arguments over and solutions to the 'urban problems' of London in the 1880s.

Changes in domestic labour

The protests among many middle-class reformers over the triviality of domestic work were not all directed at aiding women's escape from the home. Around the turn of the century proponents of the new study of 'domestic science' attempted to elevate the status of domestic work by emphasizing both its scientific and rational basis and its social value. They suggested that manual domestic work using new 'scientific' household equipment was no longer a demeaning activity for the middle-class woman. These changed attitudes both eased and arose from the adjustment of the middle classes to the declining supply of domestic servants. The fall in supply was particularly acute in the suburbs since their low density and lack of local entertainment offered few attractions to 'live-in' domestic servants. 'Daily' help was also difficult to attract, for the social homogeneity of each suburb meant that such servants now had to commute long distances to work (Miller, 1983). By the 1920s 'live-in' domestic servants were a prerogative of only the wealthier middle-class families and home ownership was replacing servants as a sign of social status. The increasing demand for labour-saving devices for cooking and housework was encouraged by manufacturers, while the new professional social workers, health officials and teachers were also anxious to instruct the working-class woman on the 'proper' care of her family. The standards of domestic work and child care were set increasingly by outside 'experts' and their achievement was dependent upon the purchase of manufactured goods: better cleaning products and tools, refrigerators, 'educational' toys and books. Thus the home, although meant to be a refuge from the outside work of commerce, was becoming its major market.

The twentieth-century suburbs

The stage was now set for the rapid inter- and post-war expansion of new suburbs, new estates, and, in Britain, of new towns. These embodied the new conception of the 'ideal' home and community in which neighbourhoods of single-family, low density housing, equipped with modern kitchens, would be clustered around schools and shops. The standard of living of both middle- and working-class women in these homes was undoubtedly better than that of their early nineteenth-century counterparts. Working-class men and women had fought successfully to improve their housing, but they had also fought to create in their home a place where they could live some part of their lives away from the demands of wage-work. Ironically their very success created a greater dependency on the 'family' wage. The income to support a 'non-working' wife and children, and the payments for the house and its equipment could only be supplied through regular employment. Thus waged work had become the key to the desired way of life, and despite much rhetoric about the virtues of women's domestic work it became thought of, as it still is today, as less demanding and less significant than paid work outside the home.

After the Second World War the problems of the isolated and restricted life of the suburban housewife became a popular topic both in the mass media and for academic study. The desirability of the social homogeneity of individual suburbs was questioned and there were debates about the size of an 'ideal' community, its social mix, its physical form, transport and service structure. Geographers as well as sociologists and planners contributed to these debates. Despite this considerable body of work few people questioned the assumption that women should have the primary domestic role, nor did they examine in any depth the interactions between the physical form and location of suburbs and women's ability to combine wage-work and family care, as many were now attempting to do. Indeed, in most of the studies of the early 1960s it was assumed that women's paid work would remain of secondary importance in their lives, that with rising affluence all women would soon have cars, and that therefore any problems of access would vanish. The solutions to women's isolation and boredom were seen to lie in improving their opportunities for voluntary work and community involvement.

We have shown that the development of a separation between home and waged work was linked both to a division of labour between female domestic workers and male wage workers and pressure to improve the supply and quality of industrial labour. The 'urban problems' and the 'woman question' of the nineteenth century resulted in state intervention to aid the process of suburbanization. State intervention has been even more important in creating the new residential environments of the twentieth century through policies affecting both planned housing and the family.

In the next section we shall look at a particular British urban policy – the building of new towns after the Second World War – in order to show how the changing relationship between men's and women's lives was reflected in, and affected by, the built environment. Finally we shall try to answer the question which perhaps you are already beginning to ask – what would a feminist, or rather a non-sexist, city look like?

Post-war new town planning: a missed opportunity to change women's lives?

Although the origins of the post-war new towns lie in Ebenezer Howard's socialist view of a more equal, and less urban, life in his community-owned satellite towns, in Britain the major impetus for large-scale development was the Second World War. In wartime, large cities were seen as a strategic danger, while after the war housing shortages were acute.

During the war, there had been radical changes in the ideology and actuality of women's role and their daily lives. Large numbers of women, married and single, entered the labour market because their contribution was needed by the state for 'the war effort'. In many cases they took over traditional male areas of employment, challenging the myth that women's only place is in the home. Large numbers of women were compulsorily drafted into war with the exception of those who were pregnant or who had children under fourteen. However, many mothers did engage in paid employment and many others undertook voluntary work. In addition, the traditional nuclear family was disrupted. Not only were large numbers of husbands and fathers absent for long periods, but from many towns and cities several thousand children were evacuated to safer rural areas. Thus the central place of mothers in child care was challenged. During the early 1940s, there was also communal provision of previously privatized, domestic services. As well as the familiar and well-documented expansion of state-provided child care, some women were freed from an additional area of domestic labour by the establishment of restaurants and cafeterias run by local authorities. At their peak in December 1943, there were just over 2000 of these. As Roberts (1981) has pointed out, these 'British Restaurants' were part of a communal feeding programme – which also included the school meals service, industrial canteens, and the rural pie-scheme for land workers – that challenged 'the assumption of women's domestic role and the supremacy of the family as the base institution of social life' (Roberts, 1981, p. 1). It might be thought that this programme would have led to changes in women's roles in post-war society and in house building and planning policy but such changes were very limited.

In certain documents of the war years and early post-war period, there was indeed some recognition of the constrictions on women's lives imposed by housing form and location. The Reith Committee Report (1946), which pre-dated the 1946 New Towns Act, for example, contained the surprisingly radical recommendation for day and night nurseries and communal restaurants. As Reith wrote: 'War-time experiences have strengthened the impulse to escape from the necessity of preparing and clearing up every meal in the week' (Reith Committee Report, 1946, p. 42), but any impression that he was concerned with equality for women was marred by his rider: 'hired domestic help is unlikely ever to become as plentiful as it once was, and women naturally want to take what respite they can from work in the home' (1946, p. 42). It seems clear that women's general oppression was only recognized because of the plight of servantless middle-class women. It is also apparent that Reith took for granted the fact that work in the home was the responsibility solely of women.

The stultifying and restrictive nature of individual domestic labour performed by women within the nuclear family had earlier been recognized. Ruth Durant in her 1939 assessment of social relations in Watling, while approving of the family in general, despaired of the 'family-centredness' engendered by the nuclear family and wrote disapprovingly that 'nowadays domesticity predominates'. To counteract this, she recommended the provision of community buildings. Two years later, the authors of *When We Build Again* (1941), published for the Bournville Village Trust, had a vision of light and fresh air, housing built around greens, with communal laundries and restaurants for 'the continued liberation of women set free from traditional constraints by war-time mobilization'. Beveridge, too, although concerned with social insurance schemes rather than urban planning, was aware of the relationship between housing form and women's oppression. Thus the Beveridge Report (1942) contained the following statements:

> The housewife's job with a large family is frankly impossible and will remain so, unless some of what now has to be done separately in every home – washing all clothes, cooking every meal, being in charge of every child every moment when it is not in school – can be done communally outside the home.
>
> (p. 264)

> Nothing short of a revolution in housing would give the working housewife the equivalent of the two hours of additional leisure a day on five days a week that has come to the wage earner in the past seventy years.
>
> (p. 275)

But like Reith, Beveridge was not quite the radical advocate of change in women's domestic roles that he at first appeared. Married women, in particular, were sternly warned that 'the attitude of the housewife to gainful employment outside the home is not and should not be that of a single woman, she has other duties. . .' (Beveridge Report, 1942, p. 52).

Communal provision for domestic labour was not, in fact, a new idea in the 1940s, for it had an honourable precedent among the founding fathers of the new towns. Unwin, influenced by American ideas, advocated co-operative housing arrangements in his 1901 publication *The Art of Building a Home*, and he seems to have interested Ebenezer Howard in the idea. In the early garden city of Letchworth, for example, thirty-two kitchenless houses with a communal dining hall were built in 1909, and further units were erected in 1915 and 1924. Initially, the *women*(!) tenants cooked one meal a day on a two-week rotation, but later a cook was employed. This system lasted until the end of the 1940s. Similar arrangements were also organized in Welwyn Garden City. Unfortunately these experiments did not influence the later generation of new town designers and builders. Concern for community participation to involve all citizens, that had so absorbed Reith, degenerated into a purely physical arrangement – the neighbourhood principle. Neighbourhoods were designed as self-contained groups of several hundred houses with associated local facilities of shops, parks, a primary school, and primary health care, interpreted by their male architects as reducing travelling time and costs for women and children, but also reducing choice and access to employment opportunities. Community provision tended to stop at a local hall provided by the Development Corporation and a pub built by one of the major brewers. Social facilities specifically for women were not provided, nor were the communal laundries or restaurants that would have reduced individual women's domestic work. Housing provision was uniformly traditional, comprising two- and three-bedroomed units, which firmly placed each individual housewife where – increasingly during the 1950s – she was considered to belong. The master plans, research reports and memoranda, and academic and official evaluations of the early new towns all ignore questions of gender differentiation, of women's waged and domestic labour, apart from a few asides about the need for locally based female employment opportunities for the 'less mobile' female population (see Hall *et al.*, 1973). Domestic virtues of 'neatness and tidiness' were reflected in the architecture and settlement design, and planners regarded themselves as 'societal housekeepers' (Hall *et al.*, 1973, p. 370). The planners' and architects' conception of gender roles was given physical expression *par excellence* in the commissioned sculpture by Henry Moore in Harlow's central square – a male hand protectively resting on a woman's shoulder as she cradles their child (Figure 3).

Figure 3 Statue in Harlow's central square by Henry Moore.

The design, management and allocation of state housing also reflects the patriarchal assumptions of local state policy. In the early post-war years there was a great emphasis on providing labour-saving devices within the home. Fitted kitchens were seen as particularly important, by the new Labour Government and by professionals alike. Dr Jane Drew, an architect, in an interview for *Women's Illustrated*, voiced this opinion: 'I feel that every woman agrees that household drudgery must be banished after the war and that's why I'm concentrating on kitchens' (in Roberts, 1981, p. 7). This policy of providing a fitted kitchen in each individual dwelling unit built by the state eventually became enshrined in the Parker Morris Committee Report (1961). While such policies certainly reduced the physical strain of domestic work for working-class women, they did nothing to reduce their burden of *individual* responsibility for that work.

The allocation of council housing also is based on assumptions about the proper tasks of women within their own homes. Prospective tenants of council houses are inspected by housing visitors (Ungerson, 1971) who may arrive unannounced, and families are graded according to their suitability for particular types of property. Considerable emphasis is placed on domestic virtues, such as housekeeping standards and cleanliness, both before and after a tenancy is allocated. In the early post-war years, explicit advice on the standards expected in the new house was handed out, without a doubt aimed at women who, virtually without exception, were excluded from the actual tenancy agreement. In the 1950s, one housing manager advised:

> Keep your home clean and tidy. Endeavour to have some method of cleaning as you go along; do not try to clean the whole house in one day. Regular bedtimes for children and adults except on special occasions. Sit down properly at table. Hang up your pots and pans or put them on a shelf. . .
>
> (in Ward, 1974, p. 12)

Such patronising advice is no longer usual, although, as Tucker (1966) makes clear, local housing authorities:

> . . .prize above all a good (i.e. solvent, tractable, clean and quiet) tenant and tend to favour *him* as any private landlord would. Because *he* is deemed likely to treat it carefully, he is generally given one of the authority's newest and best homes.
>
> (1966, p. 11, our emphasis – yet who was assumed to be doing the work?!)

By this brief analysis of new town planning, we hope that we have shown how an interest in the assumptions about women's roles provides new insights into an understanding of urban spatial structure and planning principles. While numerous commentators have pointed out the anti-urban biases in British planning which have, in part, resulted in the adoption of green belts, the neighbourhood principle, and the new towns themselves; most have failed to recognize the patriarchal basis of land use planning and the implications for relations between men and women in towns and cities.

Towards a non-sexist built environment

One of the most potent visions of a world based on co-operation rather than on competition between men and women, where domestic tasks are shared equally, is Marge Piercy's novel *Women on the Edge of Time* (1978). Such is the lack of gender differentiation that all child care is shared. However, hers is as yet a science fiction world. Nevertheless, practical experiments in creating physical environments for a non-sexist world have been tried and ideas on the form of a non-sexist city canvassed.

Members of the Owenite movement, which started in England during the early 1800s, saw the creation of new forms of settlement and building design as an integral part of their attempts to achieve equality for women and to create a more collectivized form of family life (Taylor, 1983). At the same period Charles Fourier in France also proposed communal housing and improved settlement design as essential to the creation of a society in which gender and class inequalities would be eliminated. Dolores Hayden in a book called *The Grand Domestic Revolution* (1980) has traced a long and hidden history of such ideas about housing and neighbourhood design which were intended to liberate women from isolated home-based domestic work. A variety of proposals ranging from comprehensive ideas such as those of Owen and Fourier to more limited changes such as kitchenless houses, communal centres for cooking, eating, laundry work and child care were not only put forward but also implemented in a number of experimental schemes in both Europe and America in the nineteenth and early twentieth centuries. Hayden estimates that in America about five thousand women and men had participated in experiments to socialize domestic work between 1857 and 1917. These ideas and experiments influenced Ebenezer Howard and Unwin in Britain as we have mentioned earlier (see p. 60). However, most of the experiments Hayden reports sadly were short-lived and they now seem a utopian dream in Britain and North America, both with governments wedded to traditional views of women's roles and a culture which emphasizes individualism and the privacy of the nuclear family.

There are, however, still many experiments in co-operative, non-family living being tried in different parts of the world. Some of these are based on the exclusion of men (Ettore, 1978; Keller, 1981), while others are based on a more equal division of domestic and productive tasks. Often those involved also attempt to be economically self-sufficient, turning their backs on the market. These experiments, however, tend to be small-scale and, unlike their earlier counterparts, are usually based on the adaptation of the existing housing stock rather than on new forms of building. They hardly provide a major challenge to the current operation of the land and employment markets.

At the present time there are two trends which may in part challenge the current divisions: unemployment and the electronic revolution. In households where there is no waged worker, or where traditional roles have been reversed and women, rather than men, are able to find waged work more easily, there may be some breakdown of traditional gender roles. The electronic revolution, on the other hand, may mean that certain categories of white-collar workers will be able to work from home eventually, thus breaking down the spatial separation of home and work and allowing the public domain to invade the private. However, current research on unemployed families indicates that women continue to perform the majority of domestic tasks, and the electronic revolution is still in its infancy. The ideology of home as a haven, and the assumption that domestic work and child care are women's work, is so strong that it seems to us unlikely that radical changes will occur in the gender division of labour without the creation of a new type of built environment.

What should this new built environment be like? Feminist geographers, planners and architects must develop answers to this question through both theoretical discussion and practical experiments, following the example of our nineteenth-century predecessors. However, we need to question whether their prescriptions for a non-sexist built environment are still appropriate today. Their concern was to develop more communal forms of society and to reduce the physical effort of domestic work as well as its sexist organization. To this end they hoped to harness the most up to date technology of their time. They saw in the new industrial technology an opportunity for liberation if only women could control its use. During the twentieth century this technology has been used by capital to produce domestic appliances and goods which *have* reduced much of the effort required in laundry work, cleaning and cooking. For the affluent, particularly in North America, relatively cheap 'fast food' restaurants and home delivery of prepared food offer market alternatives to the communal kitchens of the early reformers. However, these developments have *not* been controlled by women nor have they been used to reduce women's burden of responsibility for the care of their families, to change the sexist division of domestic tasks, or to promote a less individualistic and competitive way of living. If we follow the spirit, rather than the letter, of earlier attempts to create non-sexist built environments we must consider how we might use the technology of our own day to achieve their general objectives. In their day the sheer time and effort needed to perform many domestic tasks was a major obstacle to improving women's opportunities. As is made clear in the following chapters which discuss women's employment and access to facilities, the demands of child care and care for the elderly are crucial obstacles to improving women's position both at home and in the labour force.

Homebased care for children and elderly relatives is still primarily a female responsibility and, by confining women to the home for several hours of the day, these duties reinforce the allocation of *other* domestic tasks to women.

The demands for such care create so many difficulties of scheduling their activities in both time and space that few women are able to combine waged work and home responsibilities satisfactorily (Pickup, 1983). These problems of scheduling cannot be solved merely by equal sharing of domestic tasks between men and women – however desirable that may be – since the present spatial and temporal organization of urban areas hinders the integration of public and private activities for women and for men. Short-term improvements to women's opportunities could be made by increasing the current meagre provision of state child care and homes for the elderly. In the longer term, however, we need to develop new and better forms of social and spatial organization, including better forms of communal care than exist at present. We must find answers not only to such questions as, 'how and by whom should care be provided?'; but also to other questions such as, 'in what size and type of building?' and 'where should these buildings be located in relation to other land uses?'. This is a new debate in which we hope you will take part. Possibilities which could be explored are co-operative housing, 'service' units where socialized child care and other activities might be provided or shared, and the greater integration of housing for a wide range of age groups and family types.

We believe that attacking the conventional separation between public and private space both in theory and in practice should become a feminist and socialist priority in the years ahead: women cannot improve their overall position in society until their status both at home and in the labour market is altered, and this requires the breakdown of the current social and spatial division of labour and the creation of a new form of urban built environment.

Further reading

1 For an introduction to the work of sociologists and historians on the separation of home and workplace and on changes in women's employment during the nineteenth century, we recommend the readings in Part One of E. Whitelegg, *et al.* (eds.), *The changing experience of women* (Martin Robertson, 1982). A useful companion to this volume is C. Hall and S. Himmelweit, *Public and private spheres*, Open University Course U221, Unit 8 (1981), which discusses the articles in the former book. For a geographical perspective on these changes we recommend S. Mackenzie, and D. Rose, 'Industrial change, the domestic economy and home life', in J. Anderson, S. Duncan and R. Hudson (eds.), *Redundant Spaces: Social*

change and industrial decline in cities and regions (Academic Press, London, 1983) which brings out some of the important theoretical issues.

2 A seminal article on links between family life and community ideals is L. Davidoff, L'Esperance and H. Newby, 'Landscape with figures: home and community in English society' in J. Mitchell and A. Oakley (eds.), *The Rights and wrongs of women*, (Penguin Books, Harmondsworth, 1976). The article by L. McDowell, 'Towards an understanding of the gender division of urban space', *Environment and Planning D: Society and Space*, vol. 1 (1983), pp. 59–72, on which part of our discussion in this section was based, not only examines the development of British urban and housing policy in relation to gender divisions but provides a comprehensive reference list and brief review of literature on women and urban structure and change. Our discussion on p. 46 also gives some guidance on feminist work on current urban issues.

3 Lively and interesting accounts of earlier ideas on non-sexist urban environments can be found in D. Hayden, *The Grand Domestic Revolution* (MIT Press, Cambridge, Mass. 1980), and in B. Taylor, *Eve and the New Jerusalem* (Virago, London, 1983).

4 Information on data sources for urban studies can be found in J. Short, *Urban Data Sources* (Butterworths, London, 1980).

Topics for discussion

1 To what extent does the spatial separation of women into the 'private sphere' facilitate or hinder the development of community action by women to improve their local environment?

2 How far do class divisions cut across gender divisions in western capitalist cities?

3 Discuss some of the major ways in which gender divisions in cities might be reduced.

4 Women's employment, industrial location and regional change

Introduction

Since the end of the Second World War major changes have taken place in the composition of the labour force in Britain. Firstly, the overall number of women going out to work has increased dramatically. Secondly, there has been a significant separation of the type of work done by women and that done by men. Thirdly, a distinct inter- and intra-regional distribution of 'jobs for women' has evolved, especially since the 1960s. Finally, many women have lost their jobs in traditional sectors of female employment, such as textiles, while others have gained them in the 'new industries' of electronic and electrical engineering or in the service sector. Such processes of change have not, until recently, been the focus of either research or teaching in geography. Nevertheless they are important, since these trends have not only affected the everyday lives of women and their families, but have also influenced the ways in which goods and services are produced.

Changes in the patterns of women's employment have taken place alongside increased state intervention in the economy and in the physical environment. Economic, regional and urban planning have variously influenced both the physical structure of our towns and the social fabric of post-war Britain. But investment policy and location decisions for housing, transport, education and other 'social facilities', as well as industry, have assumed that women are primarily mothers and housewives who remain in the residential environment (Women and Planning Group, 1982). Women's employment has been regarded as marginal and therefore less important in the development of either prosperous local economies or of a 'healthy' Britain as a whole. Meanwhile changes in the manufacturing processes of industrial firms and the organization of work in offices, shops, and other institutions such as hospitals and banks, have made demands for different workforces, including large numbers of women.

Current empirical research on industrial location and regional inequality clearly points to the links between women's changing employment patterns and the processes of economic change. Throughout this chapter we will draw on two recent studies of women's employment and home lives in the North-East of England and West Central Scotland to help illustrate these developments. Part of this work has concentrated on new towns where the growth in female employment has come into conflict with the ideology which has governed urban spatial structure (see Chapter 3, p. 59) Jane Lewis's study of Peterlee found that unusual emphasis was placed on the need to attract firms which would specifically employ women. In a similar study of East Kilbride, however, Jo Foord discovered that the dramatic increase in women's employment was a 'by-product' of growth sector policy directed at attracting new male employing industries to greenfield sites.

However, we will first show where and in which sectors of the economy women's employment has grown. We will also ask why women's role in the labour market has been missing from regional analysis, and so present new arguments for understanding the geography of women's employment.

Women are commonly seen first as 'domestic labourers' and only second as 'waged employees'. Contrary to this assumption, most women go out to work to support children and make an essential contribution to the household budget. The composition of the family is changing, the proportion of single parent families and cohabitees has grown. Only 5 per cent of all households in Britain live in the traditional family group with a male breadwinner, female housewife and two dependent children (Coote and Campbell, 1982). Currently, one in eight children live in single parent families. Such families are predominantly lone mothers with their children and evidence suggests that when these women go out to work, as many of them do, they work full-time as the sole supporter of the family (Rimner and Popay, 1982).

Nevertheless, the view that women's waged work is secondary and so insignificant as a major source of income, is persistent. It helps both to maintain women's low wages and unequal job opportunities and to undermine women's own expectations of waged work. Given the prevalence of this view, it is hardly surprising that conventional geography has neglected women's employment, but since many women do combine their role as housewives and mothers with waged employment there are several new questions with which geographers should be concerned. Not only should we ask where and at what are women working, but also how they combine these diverse activities both in time and space. Have they adequate transport and child care facilities, and if so, are such social resources located to maximize efficiency for *women* or for industry? In the final section of this chapter we will discuss some of these questions in relation to women's own experience of employment.

Women go out to work: a new geography of employment

The number of women taking on jobs outside the home has steadily increased throughout the post-war period. In 1951, women made up 31 per cent of the total British workforce. But by 1978 they had reached a peak of 42 per cent. The number of women in employment increased by 48 per cent between 1951 and 1981, to a total of 10.2 million. Meanwhile the number of men in employment decreased by about 2 per cent. More significantly, however, this female increase was largely made up of married women. In 1931, for example, only 4 per cent of the total workforce was accounted for by married women. By 1981 this had risen by over six times to 26 per cent. In 1982, of the approximately 10.2 million women who were economically active, 65 per cent were married. In addition, many of these women had young children. In 1981, 21 per cent of all mothers with children under four years old and 46 per cent with children aged between five and ten had a paid job. The percentage of mothers who worked for wages rose from 15 per cent in 1951 to 54 per cent in 1980. Hence, most women who have entered the workforce have also taken on the double burden of home responsibilities and the demands of waged employment (Aldred, 1981; Low Pay Unit, 1979).

These total figures conceal another significant aspect of this massive growth. Women have moved overwhelmingly into the service sector. This means they do jobs in shops and offices; in hotels, bars and restaurants; in schools and hospitals; or in warehouses from which finished goods are distributed. Approximately 75 per cent of all women workers are in this sector. However, many women still work in the manufacturing industries; but this is a shrinking source of employment for both men and women. Between 1961 and 1980 manufacturing industries in Britain lost 2.3 million jobs of which 27 per cent were lost to women, while the service sector gained 2.5 million jobs, 84 per cent going to women workers (Coote and Kellner, 1980; West, 1980). Only a few manufacturing industries have large female workforces. These are the clothing and footwear industry, the textile industry, the food, drink and tobacco industries, and the electrical engineering and electronics industries. In fact 50 per cent of women working in manufacturing are employed in these four, out of a total of seventeen, manufacturing groups. On the other hand, no one industry employs as many as 10 per cent of the total male industrial workforce (Hakim, 1979; Oakley, 1981).

So far we have shown that there has not only been a large number of women entering the labour force, but that they are mainly married women and that they work either in a few manufacturing industries or the expanding service sector. However a gender division of labour in waged work is not only visible between sectors and industries. Surveys of employment status show clearly that women occupy the lowest skilled jobs which require little official training and return the lowest pay (Hakim, 1981; Philips and Taylor, 1980). It is often repetitive work and either requires the 'dexterity' to assemble components quickly in a factory or the patience to carry out routine clerical and cleaning tasks in office, hospital or shop. It also includes work catagorized as 'personal services' such as hairdressing, auxiliary nursing and child-minding. Segregation is very clear cut. Whereas men dominate in managerial and professional occupations, there are four times as many women as men in clerical or equally junior non-manual positions. Similarly very few women are classified as skilled manual workers. They work predominantly in jobs classed as semi-skilled.

Women have not therefore entered the labour force on an equal footing with men. The Peterlee Study found that during the 1960s and 1970s, when the bulk of women took on employment outside the home, the trend in manufacturing was towards employing unskilled women in assembly plants and employing men, rather than women, in higher graded skilled or professional occupations. Changes in production towards mass assembly and rapid technical and professional specialization in industry accelerated this gender separation. Similar processes were at work in the non-manual occupations in East Kilbride, with women being recruited into the more junior clerical positions.

In general, job insecurity and few fringe benefits accompany women's low rates of pay. On average full-time women workers earn only 64 per cent of the male wage packet. In 1981, 20 per cent of manual and 7 per cent of non-manual full-time women workers earned less than £60 per week. But less than 1 per cent of both manual and non-manual male workers worked for equally low pay (Low Pay Unit, 1980).

Discrepancies between men's and women's weekly earnings are paralleled by differences in wage levels between employment sectors. Where male employees dominate, wages for both men and women tend to be higher (i.e. in the petroleum and chemical industries). But where women employees dominate, the reverse is true and wages for both men and women are much lower (i.e. in the food, drink and tobacco industries, the clothing industry, the electrical engineering industry and the distributive trades).

However, average wage rates disguise several other factors which strongly influence women's earning capacity. The two studies found that women were in the lower graded occupations, they did not do as many hours overtime as men, rarely worked night shifts and did not benefit so often from extra bonuses or fringe perks (i.e. pension schemes, use of firm's van or car). Also a large number of the women surveyed worked part-time and so were excluded from overtime and bonuses altogether. Indeed, low pay and part-time hours went hand in hand. The gap, therefore, between individual women's and men's earnings is, if anything, wider than generally assumed.

We have argued earlier in this book that women's inferior social position is related to the patriarchal organization of reproduction. The inequality in employment patterns and opportunity documented above is similarly linked to patriarchal relations inherent in both the home life and workplace experience of women employees. The difficulties that face women in combining their 'dual roles' of waged employment and domestic and child care responsibilities do restrict women's choice of jobs. Their opportunities are also limited by the time women spend away from work bearing and looking after children.

In Peterlee, women often only looked for work in locations within walking distance or an easy journey by public transport. Also they were frequently restricted to jobs which made the minimum demands or least conflicts with their home lives. This is regardless of these women's own desire for challenging employment or career fulfilment. A lack of training opportunities or an inability to enter higher education further hindered women in East Kilbride. The 'dual role' not only limits women's ability to search for or take up a wider range of jobs, it often means they are confined to lower grade occupations and/or have to interrupt their working lives to care for children, sick people or elderly relatives. In a national survey, it was found that 50 per cent of housewives aged between 35 and 64 years old provided some care for elderly or infirm people (Hunt, 1964). Another 'solution' to the 'dual role' is often found in part-time and shift work. During 1981, while 41 per cent of all married women workers worked 24 hours per week or less, 70 per cent of working mothers with dependent children worked part-time. Of working mothers with dependent children, 75 per cent had a youngest child under the age of five years. The age of the women's youngest child is therefore the principle factor in determining whether or not she works part-time (Rimner and Popay, 1982).

Low pay and part-time hours go hand in hand. The average part time weekly wage for manual women workers in 1978 was £30.90 a week, and for non-manual £42.70. This is frequently the only source of earned income in a household, and barely maintains life at the breadline (Low Pay Unit, 1978).

Regional variations in 'jobs for women'

So far we have documented the expansion in and unequal nature of women's employment on a national level. However there are additional inter- and intra-regional aspects to this pattern of social change. Consequently a distinct geography of women's employment has emerged since the war.

The growth of women's job opportunities, which was most apparent during the 1960s and 1970s, varied both between and within the different regions of Britain. Women have formed a flexible pool of workers with the qualities modern industry required (Bruegal, 1979). The introduction of new assembly techniques and labour intensive service activities both demanded new sources of labour. Women proved relatively cheap to employ, often inexperienced in waged work or union matters and available in large numbers. The traditional

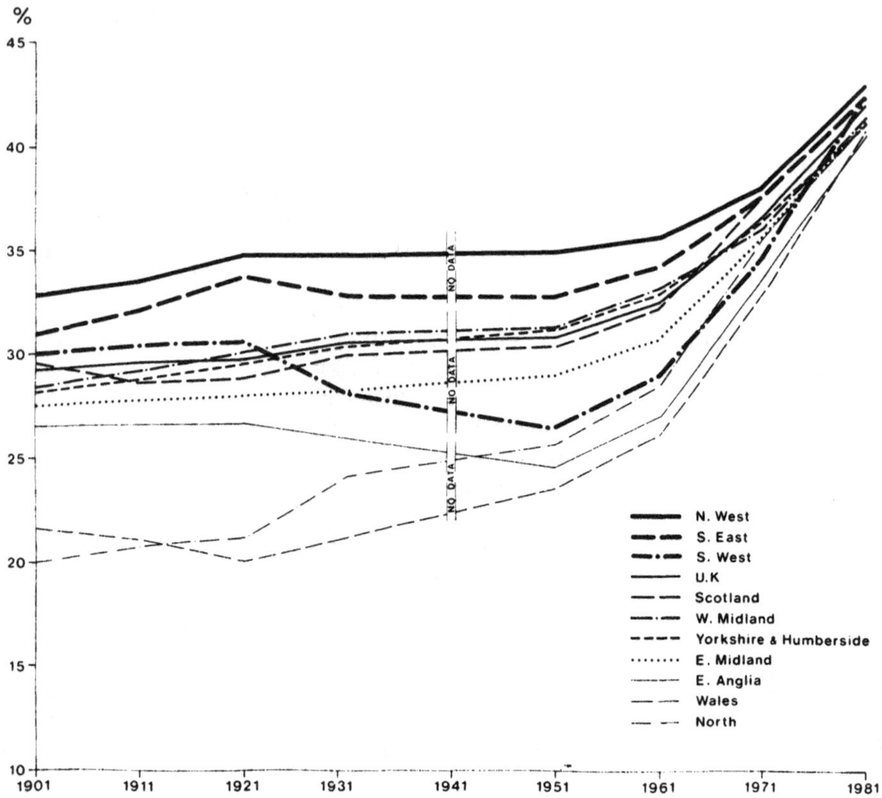

Figure 4 Female employees as a percentage of total employees by Standard Region, 1901–81

female attributes of 'dexterity' and 'docility', which girls often learn at home in the nuclear family or at school, were quickly adapted to suit the workplace. But the availability of labour with such desirable characteristics varied between the regions.

From Figure 4 it is apparent that in areas where female participation in the labour force was low before the war, a sharp increase occurred after 1961. These areas were either the older industrial regions where male employment in shipbuilding, heavy engineering and mining once predominated, as in Scotland; or the rural areas of the South-West and East Anglia. Wales and the North combined both traditional industries and agriculture and also showed rapid increases in women's employment. Both Peterlee and East Kilbride are included in these regions of rapid expansion in women's waged work.

The regions in which the rate of increase was relatively small were the prosperous areas of the South-East and West Midlands, both with a history of a high demand for labour, and the North-West, with a long tradition of female employment in the textile industry. However these distinct regional variations in the proportion of women in local labour forces have not resulted solely from absolute declines in the number of men in work. A fall in male employment has occurred but this has been small in comparison with the increase in female workers. Consequently, this new geography of female employment has been influenced both by pre-war patterns of women's work and by post-war changes in industrial production.

Data on regional changes between 1966 and 1978 in both the number of men and women working and their participation rates can best be described by identifying four distinct types of regions (Figure 5).

Within these general trends for each region other patterns emerge for changes in female employment by sector. Dominance of manufacturing growth in the Intermediate Regions is evident, as is the above average increases in service employment in the growth regions and the South-East. Such regional aggregate figures do not, however, show the type of work lost and gained in each sector. Research has shown that the above average expansion in female employment in the peripheral regions have been in routine clerical jobs in both the private and public sector, and in semi-skilled manual work in branch manufacturing plants of large national or multinational firms (Firn, 1975; Watt, 1982).

This inter-regional change in the nature and location of women's employment is, however, only one of the processes underlying the new pattern in the geography of women's employment. Between 1951 and 1971 there was also a marked decentralization from larger urban areas in the overall distribution of female employment. This loss accelerated in the ten years between 1965 and 1975 coinciding with the rise of inner city job loss, the decline of traditional manufacturing and the growth of assembly and service sector industries in the suburban and peripheral areas of metropolitan boroughs (Corkindale, 1980).

These general patterns of inter- and intra-regional employment change also disguise the locational concentration of women's waged work within particular local areas. The gender recomposition of employment has involved a geographical reorganization of local labour markets. Thus, for example, in the North-East, new female-employing firms are concentrated in the new towns at Peterlee, Washington and Newton Aycliffe and in the post-war industrial estates on the periphery of Sunderland, Newcastle and the Hartlepools. The reorganization of retail and distribution industries and of government

Figure 5 Regional employment change, 1966–78 (See text for explanation)

1 'Growth Regions', in which both male and female employment increased by more than the national average. East Anglia and the South-West and the East Midlands fall into this group.
2 'Intermediate Regions', where total employment change is close to the national average because an above average loss in male employment has been off set by an above average gain in work for women. These changes have been seen in the northern and western peripheral regions – Scotland, the North and Wales.
3 'Declining Regions', in which both male and female employment loss is greater than the national average change. Such areas are in the North-West, West Midlands and Yorkshire-Humberside regions.
4 'The South-East', in which both male and female employment changes differ little from the national average. The South-East includes Greater London and the Home Counties.

(Bowlby *et al.*, 1983)

departments has led to a similar concentration of female service sector jobs in West-Central Scotland.

The large entry of women into the local labour force changes the nature and composition of the working population. It also alters the aspirations and needs of individual workers. New issues arise with different patterns of work, part-time hours and 'twilight shifts' (evening hours, often from 5.30 to 10.30). Child care issues and maternity leave, as well as demands for better pay and conditions of work, accompanied women's waged employment. But the traditional local and national trades union organization has often found it difficult to support and fight for women's rights at work, or indeed women's rights to waged employment at all (Boston, 1980; Brietenbach, 1982; Cockburn, 1983; Gregory, 1982). Men's rights to a 'family wage', enough to support a wife and children, are central to traditional wage bargaining in Britain but the antithesis of women's demands for equal pay. This reflects and subsequently reinforces patriarchal forms of social organization. For example, an equal pay struggle at a clothing firm in East Kilbride during 1975 was defeated within the union before reaching a tribunal.

Similarly, gains in women's work in areas of high male unemployment have been seen as second-best and at the expense of 'real jobs' for men. Part of this hostility to women's employment in local areas stems from the threat it poses to the traditional image of male and female roles. In regions previously dominated by male employment, loss of work for men and loss also of their breadwinning role has been further compounded by the availability of waged work for women who have traditionally stayed at home. This opposition does, however, have a material basis. Loss of higher paid skilled male employment and gains in lower paid female employment have been seen as a threat to the projected economic viability of local areas.

The geography of women's employment: a review of the literature and some suggestions for a way forward

So far we have outlined the changes in the quantity, location and nature of female employment in post-war Britain. This section will now consider some questions arising from this discussion.

Firstly, why should an understanding of the position of women in the labour market be a central concern of regional analysis? The geography of female employment growth does have significant implications for new patterns and forms of regional inequality in Britain. While the change in the gender composition of employment has been a national phenomenon, it has been accentuated in the older industrial areas of Wales, the North and Scotland. Furthermore, this change in the gender composition of the workforce has also meant a change in the gender of those excluded from employment. Meanwhile, the nature of work itself is undergoing a related transformation.

It has been increasingly recognized that the creation of new jobs in itself is not a sufficient guarantee of future economic development in depressed areas. It may well be that the low skill and low wage structure of many of the new jobs, done by women, in these regions have perpetuated or even exacerbated existing patterns of inequality (Austrin and Beynon, 1979; McNabb, 1980; Massey, 1979; Perrons, 1980). Their creation has certainly altered the form of regional development.

Changes in the organization and location of industry have generated a demand for female labour. But these changes have also been made in response to, and have been shaped by, the existence and specific geographical distribution of female labour supplies. Without such female labour reserves, the reorganization and relocation of industry may well have been very different. The particular location of female labour reserves was in part a result of previous patterns of industry and employment. In the North-East and West of Scotland, for example, the former dominance of traditional industries employing a predominantly male workforce, the nature and status of male employment which placed heavy demands on female domestic labour in the home, and the lack of alternative employment opportunities for women firmly entrenched a strict division of labour between men as breadwinners and women as housewives and mothers. These were all important elements in producing the large reserves of female labour in these regions and in providing growth industries with a necessary new source of labour. Moreover, it was not only the numbers of women available for wage-work which attracted new investment to these areas, but also the *type* of female labour available. The Peterlee Study found that older, married women were often regarded by employers as both a more reliable and more flexible source of labour than younger or unmarried women. Many of these women had no experience of factory work, nor, indeed, of working for a wage at all. The female labour reserve in this older industrial area therefore provided mobile manufacturing industry with a 'green' and a relatively cheap new source of labour. Concentration of female labour reserves in a few localities has also attracted private and public sector employers. In the East Kilbride Study, similar processes of reorganization, recomposition and relocation were identified in the distributive trades and public administration.

It has recently been argued that regional labour reserves have occupied a vital role in the post-war rounds of investment, and that the state has become increasingly involved in their creation (Damette, 1980; Hudson, 1982). Throughout the post-war period state spatial policies – in the form of regional economic and new town policy – have become increasingly involved in the provision of social infrastructure and freely available housing. Such measures were undertaken especially to attract new capital investment to the depressed regions. In particular, the growth area policy dominant in regional planning in the early 1960s emphasized the efficiency of concentrating public investment on a single site. One of the major consequences of these policies has been to assemble large pools of labour (Lewis, 1982). Whereas in Peterlee such policy was deliberately associated with grouping women workers together in one locality, in East Kilbride emphasis was more generally on creating a pool of 'quality' labour, in terms of appropriate skills and low levels of militancy.

Existing patterns of regional development are therefore in part the product of former divisions of labour between men and women. Gender relations, social relations between men and women, vary between different areas of the country. Consequently the nature of and the variation in gender relations have been integrally related to the changes which have taken place in the organization and location of industry and employment. So a specific examination of the geography of female employment should be *central* to regional analysis.

Our second question is, therefore, to what extent are any of these aspects of the geography of women's employment considered in the existing literature within regional analysis? With a few exceptions (Cooke, 1981; Hudson, 1980; Massey, 1983; Williamson, 1982) there has been no analysis of the specific nature and impact of the spatially differentiated growth of female employment. In particular, there has been little attempt to explain the processes underlying aggregate regional patterns of female employment change. In general, female labour is seen as an important location factor in the geographical decentralization of manufacturing activity since the early 1960s. Much of the existing literature can be divided into four categories in terms of its treatment, or absence of any specific treatment, of the role of women in the labour market in recent processes of industrial and regional change.

First, there are studies which examine changing patterns of regional inequality by using measurements of aggregate or male employment and unemployment patterns. This has been common in analyses of the impact of state regional policy on employment in the Assisted Areas (Keeble, 1977; McCrone, 1969). The use of aggregate or male employment statistics in such studies has meant that the substantial changes which have taken place in the gender, occupation and wage composition of employment in the Assisted Areas have often been overlooked. Failure to disaggregate employment statistics by gender has given credence to the view that regional policy has been relatively successful either in terms of creating new employment to replace jobs lost in the Assisted Areas, or in terms of equalizing employment patterns between regions. But in ignoring major changes in both the type of employment created and gender of workers employed, this work distorts our understanding of processes of industrial and regional development as a whole.

Secondly, there are studies which offer explanations of the regional variations in women's economic activity by measuring correlations between regional differences in female activity rates and other assumed determinants of women's participation in the labour force. For example, variations in female activity rates have been explained as functions of regional differences in the degree to which mothers with dependent children leave the labour market (Joshi and Owen, 1981), in the extent to which there is a tradition of women's work or the amount of part-time work for women (Allin, 1982). Similar explanations as functions of regional variations in industrial structure (Bowers, 1970) or in the degree of urbanization (Greenhalgh, 1977) have also been put forward.

While each of these factors may well have a significant impact on women's participation in the wage labour force, they do not *explain why* female activity rates are regionally differentiated nor do they attempt to explain the distinct geography of female employment *growth*. While Allin (1982) does point to the importance of a tradition of women's participation in paid work in regions with high female activity rates, he overlooks the fact that the *rate of growth* of female employment has been highest in those regions where women's participation in the labour force has been traditionally low. As explanations of regional variations in female activity rates these studies are inadequate; they do little more than describe the correlations between regional variations and different aspects of women's participation in the wage labour force. We must move beyond measurements of regional patterns of female economic activity, to an analysis of the mechanisms of change through which such regional variations in women's participation in the labour force have been produced (Sayer, 1979; Urry, 1981).

The third body of literature is conceptually close to the studies discussed above. Analyses of the spatial decentralization of manufacturing activity point to the importance of female labour as a location factor. The geographical decentralization of manufacturing activity has been perceived as a response to the rise of 'agglomeration diseconomies' in metropolitan areas (Keeble, 1981). Metropolitan diseconomies include high rents and rates, deteriorating infrastructure, traffic congestion and the age and unsuitability of premises. Conversely, economies to be reaped in non-metropolitan areas include the availability of cheaper land, rents and rates, cheaper labour and government grants, and other similar incentives (Fothergill and Gudgin, 1979; Keeble, 1981). Keeble, for example, uses indices such as the density of manufacturing employment, residential preference and female activity rates as the correlates of the geographical decentralization of manufacturing industry. But in these studies 'explanations' of the spatial decentralization of manufacturing capacity are provided by describing the characteristics, and not by explaining the processes of decentralization (Miles, 1982). Consequently, there is no coherent explanation of the processes of change taking place within the specific industries, firms and plants which have relocated to non-metropolitan and peripheral locations. Furthermore, female labour reserves are seen *a priori* as relevant to an understanding of the causes of the geographical decentralization of manufacturing activity. But, the importance of female labour as a location factor cannot be understood independently of an analysis of the changes taking place within industry, which have resulted in the incorporation of large numbers of women into the labour force in particular industries, plants and localities. Yet within these studies, changes in the location of industry are seen almost wholly as a response to changes in the geographical distribution of the requirements of production; to changes in the 'locational surface', and not, as must also be the case, to changes within industry itself.

Finally, critiques of the three previous forms of explanation of the geographical reorganization of industry have been developed in marxist and radical political-economic analysis of industrial and regional development. Although little of this work has dealt explicitly with the geography of *female* employment change (with the exception of Massey (1983) and of studies of particular local labour markets such as Merseyside (Williamson, 1982) and Washington New Town (Hudson, 1980)) it has provided a framework of analysis from which we can begin to understand why the demand for female labour has not been geographically uniform. However, while marxist and radical political economic analyses have attempted to specify the mechanisms underlying recent changes in the location and labour requirements of industry, they have not usually considered the specific role of gender divisions in these developments (again Massey, 1983 is an exception). In other words, they have not asked why a demand for *female* labour was generated and why women are cheaper and considered a more flexible, dextrous and docile source of labour. To some extent this reflects a more general problem with marxist analysis. Much of this literature is concerned with historical, political and economic explanations both of developments in the labour process and of processes of capital accumulation underlying phases of regional change (Carney, 1980; Dunford, 1979; Dunford, Geddes and Perrons, 1981). But it has failed to recognize the significance of *gender* divisions (or, indeed, of any divisions) within the labour force as an integral part of the changes which have taken place in the post-war geographical reorganization of industry and employment.

The starting point of these analyses is that changes in industrial location are an essential part of wider changes in the organization of industry (in company ownership and production technology, for example). Furthermore, they claim that these developments can only be understood within the context of changes in national and international economic conditions. Individual industries and firms will respond to pressures to reorganize production – in the light of increasing international competition, for example – in a variety of ways. On the other hand, industry operates (and investment and disinvestment decisions are taken) within an environment in which conditions, or demands, of production (such as labour availability and skills, or wage levels and sophistication of union organization) are unevenly distributed in space. Changes in production and location are therefore made in response to, and are shaped by, the geographical distribution of the conditions of production which are themselves continually changing.

The inadequacy of radical approaches raises our final question. How can we begin to understand the role of gender relations in recent phases of industrial and regional development?

Within the existing literature references are commonly made to the 'suitability' of female labour. Women are seen to be particularly appropriate for light manufacturing industry – such as electrical assembly work – because of their greater 'dexterity'. Women's work is classified as semi-skilled or unskilled. Women are seen as a more flexible and more docile source of labour, and these 'characteristics' of women workers are rarely questioned. Recently, however, developments in feminist theory have begun to question the objectivity of these traditional skill classifications which work to the detriment of women workers and undervalue women's work relative to that of men's (Alexander, 1980; Mackintosh, 1981). As Massey suggests:

> It is not the fact that these people are biologically female which explains their characteristics and their new availability for the wage work force. . .but the fact that certain kinds of (patriarchal) social relations construct them as having these characteristics.
>
> (Massey, 1983, p. 86)

Traditional skill classifications disadvantage the majority of women workers because their jobs are defined as semi-skilled or unskilled. An examination of the impact of the Equal Pay Act in Britain illustrates the extent to which skill and job classifications are maintained through the relative bargaining power of employers, trade unions and employees. Surveys of the impact of equal pay legislation suggest that the extent of job segregation between men and women renders equal pay based on job comparability an irrelevance (Hakim, 1981; Snell, Glucklick and Porall, 1981). Snell *et al.* also found that job regrading and re-evaluation schemes were commonly used by employers, in collusion with the trade unions, to avoid any job comparability between men and women's work which did exist. One contemporary Guide to the Act, for employers, even suggested that jobs could be regraded in terms of 'heavy' and 'light' work to ensure that women would receive lower wage rates (Paterson and Armstrong, 1972)! It is therefore clear that definitions of skill reflect *social relations* and not biological differences.

It is commonly argued that women's specific position in the wage labour market – their concentration in lower paid jobs defined as less skilled – is a reflection of women's 'dual role' in combining wage and domestic work. This view, common within the regional literature, focuses on the conditions determining female labour supply, on women's availability for paid employment, as an explanation of women's subordinate position in the labour market. However, while accepting that the division of labour and women's role in the family are major influences on women's position in the labour market, they are not the only ones. Above all this view fails to take account of the relationship between the gender division of labour and the organization of production as a whole. The gender division of labour is both incorporated into and maintained by the organization of the production process and working practices within industry. Indeed, the organization of production is itself an active element in the reproduction of the gender division of labour. Recently there have been several case-studies of women's work, in the print industry (Cockburn, 1983), in the clothing industry (Coyle, 1982), and in tobacco processing and car components plants (Cavendish, 1982; Pollert, 1981). All of these studies emphasize how changes in the organization of production techniques and working practices shape women's specific position in wage-work. In addition, they also emphasize the vital role of male dominated trade unions, their customs and practice, in reinforcing patterns of job segregation by gender. Unfortunately, however, very little of this feminist literature recognizes the crucial significance of the geography of gender relations in the reorganization of industry and employment.

Within the electrical engineering industry, for example, pressures to remain competitive and cut production costs involved a variety of major changes in company structure, in the organization and location of production operations and in the gender composition of the labour force. During the 1960s many companies in this sector combined, thus allowing them, as a consequence of their increased size, to invest in new product development and new production technologies. The growth in the size of individual companies and the introduction of new production techniques led to changes in the industry's demand for labour and enabled companies to fragment their operations, both functionally and locationally. This meant that while research, marketing and management activities were located in Southern England, factories manufacturing and warehouses distributing the goods were sited in peripheral areas, such as Scotland and the North. Within the actual manufacturing process, the standardization of products and the introduction of mass production techniques facilitated an overall deskilling of the workforce. Increasing numbers of semi-skilled and unskilled workers, and particularly women assembly workers, gradually replaced some of the older male dominated skilled sections of the workforce. Changes in the labour requirements as part of processes of reorganization in both company structure and production techniques have therefore enabled individual firms to fragment their activities spatially. In relocating separate stages in the overall production process to areas where conditions are most appropriate for that particular function, use has been made of local reserves of female labour. The geographical distribution of different types of workforce, in terms of skills, wage levels, degrees of trade union organization and workplace militancy, are uneven. So semi-skilled and unskilled production processes have often been located in areas with an abundant supply of relatively cheap and relatively inexperienced unskilled labour. These conditions were satisfied by the availability of female labour reserves in the Development Areas of the North and West and in small and medium-sized towns in predominantly rural areas. Both Peterlee and East Kilbride fulfilled these requirements. Between 1965 and 1975, for example, East Kilbride attracted 40 per cent of the net growth in electrical engineering employment in Scotland. By 1974, 30 per cent of all women in employment in the town worked in this sector. However these establishments were largely branch manufacturing plants of major national or international companies employing women as semi-skilled operatives.

This empirical work therefore suggests that gender relations have played a significant role in shaping regional space, and thus that the role of women in regional labour markets should be central to regional analysis.

Women's employment and the recession

Most of the changes in the geography of women's employment discussed so far have been concerned with the growth of 'jobs for women' in both the outer areas of metropolitan boroughs and the peripheral regions during the 1960s and 1970s. However Britain is now in a period of economic crisis and job loss is far more frequent than employment gain. Despite the massive post-war expansion in female employment it has often been argued that women's jobs are first to go in a recession (CIS, 1978; Walby, 1983).

Both male and female unemployment has been rising since the mid-1960s. This has largely been due to loss in manufacturing. Consequently, the impact has been greater on men. Until 1978 the overall number of women's jobs was still rising with new jobs in the intermediate non-manual and junior occupations (i.e. cashiers, clerks, typists, nursery workers, library assistants, welfare workers and junior technicians). These compensated in some way for the female manufacturing loss. Nevertheless, more recently the total number of the unemployed, and especially the number of unemployed women, has increased considerably with both cuts in the public service sector, and employment loss through automation and production streamlining in private service industries as well as manufacturing (Huws, 1983). Indeed, the concentration and then subsequent closure of branch plants in Peterlee and East Kilbride has meant that local rises in the number of women's jobs have been more accentuated than national trends might suggest.

Official unemployment rates for women are, however, generally accepted as inaccurate. The majority of women looking for work are not counted in the statistics. Two out of every three married women are not entitled to unemployment benefit and as a result many do not bother to register as unemployed. Part-time workers are not allowed to register. The justification for excluding many women from the unemployment register is again based on the assumptions that, as primarily housewives and mothers, they are neither 'real workers' available for work nor actively seeking full-time employment.

Many women do stop looking for work when overall unemployment increases. Both new town studies found that women with unemployed husbands were more likely to become full-time housewives and not to seek re-employment, if they themselves lost their job, than women whose partners had work. However, very few women in these surveys actually left a job when their husbands became unemployed. The workings of the social security system, and the fact that the areas of high male unemployment are also areas where there is increasing female job loss, discouraged women generally from seeking work. Also many women who had lost their own jobs did tend to see themselves as housewives and not unemployed workers. This self re-definition means that they moved out of the job market with little notice being taken.

The beginning of the 1980s, therefore, saw a sharp deterioration in women's employment prospects. For the first time since the 1950s the total number of women employed dropped. Given the extent to which new technology in manufacturing and service employment further threatens women's work, the future is not rosy. Just as the pattern of women's employment exhibits segregation within and between industries, so do the comparative rates of job loss. Where men form the bulk of the workforce, their rate of unemployment is higher. Conversely where women are a greater part of the workforce the opposite takes place. Within given industries and sectors – for example, the clothing and footwear industry, food and drink, public administration etc. – women and particularly part-timers suffer higher rates of job loss. It is true to say that women have been drawn in and out of work more frequently than men, especially in manufacturing. This suggests that employers view women as a more 'disposable' part of the workforce (Bruegal, 1979; CIS, 1981). There are many reasons why women are more likely to lose their jobs in a period of economic recession: lack of security, poor union organization or support, lower skill levels and little commitment by national or multinational corporations to maintaining the branch plants where women work. Many women's own confused ideas about their rights to work exacerbate this situation.

The spatial pattern of women's job loss has been as distinct as the pattern of previous employment growth. The inner areas of industrial conurbations and the traditional employment blackspots have continued to lose both male and female employment. However, female rates of job loss in the South-East, in two of the 'Growing Regions' – East Midlands and East Anglia – and the recently declining West Midlands have escalated since 1978. These areas have exhibited figures hitherto only seen in the poor post-war female employment areas of Yorkshire and Humberside and the North-West.

Although these spatial dimensions to women's unemployment are part of sectoral changes in economic activity, there is no evidence to suggest that as semi-skilled industrial manual workers lose their jobs they can move wholesale into other forms of employment in the service sector. The service sector expansion is no longer fast enough, nor does it require so many low-skilled workers. Recent evidence suggests that the recession is altering further the *quality* of waged work available to women. It is also reinforcing the narrow range of occupations open to women and widening the gap between men's and women's weekly earnings. Flexibility in working hours, job mobility and the number of part-time jobs have also been curtailed, forcing women back into domesticity.

Women's ability to take on waged employment is dependent on managing their domestic duties as well. So a range of additional factors exert an important influence on whether or not individual women in particular localities can maintain the 'dual role'. Central and local government policy on resource provision for child care, nursery education and after-school training opportunities, as well as elderly and handicapped day and residential care, all influence women's job chances, as does the availability and cost of public transport. But savage cuts in local and national expenditure have, to varying degrees in each city and region, eroded the already small state provision and commitment in all these areas. Public service cuts affect women in two ways, therefore: as clients and as workers. Loss of services have been paralleled by women public service workers facing redundancy. Increased domestic responsibility has befallen a large number of individual women and this is exacerbated by poverty where women's jobs in both the public and private sectors are fast diminishing.

Women's experience of work

So far we have shown, in some detail, the facts about women's participation in the labour force. We have concluded that the nature and distribution of female employment requires a new feminist regional analysis. It also needs a different understanding of how definitions of 'skill' are allocated to different tasks, and how notions of 'work' are socially constructed in terms of gender as well as class. Only with such tools can we explain the changing geography of women's work. Part of this fresh explanation re-examines national, regional and local patterns of work to show the links between changes in women's employment and the processes of social and economic re-organization. However, it also demands an altogether different approach to intellectual enquiry.

An understanding of women's own experience of work – both paid and unpaid – is essential to any feminist analysis. But there are only a few studies of women's experience of employment and how it influences their own lives and those of their families.

Early employment attitude surveys established that women work not for 'pin-money' but to pay for basic household needs (Jephcott, 1962; Klein, 1965). But the design and interpretation of this research also assumed that women sought employment out of some benevolent extension of their domestic caring role (in providing the 'little extras' to improve the family's living standards) rather than as an escape from domestic oppression. Not until material conditions and the ideology of gender, which place women in the home, were recognized did new questions arise: questions which linked women's employment to their situation as housewives and mothers (Oakley, 1981). Women in East Kilbride clearly stated that the loneliness, monotony, isolation and the financial dependence of being at home plus a sense of social inferiority led them to seek paid work when it was available. But taking paid employment was usually dependent on organizing, often with other women, the care of children or elderly people. Consequently stress, worry and guilt were common emotions experienced by women with jobs.

Nevertheless, even in employment women still experience the social inferiority and inequality wrapped up in definitions of their domestic role. Women's concentration in low status jobs; the control exerted by the hierarchy in which male foremen and managers often oversee women workers;

unsympathetic working of maternity leave; and contradictory attitudes towards women workers (especially black women workers) by unions and management maintain their subordination. Similarly women's awareness of the definitions of appropriate dress, behaviour and language, and their vulnerability to all levels of sexual harassment adds a further aspect to that subordination (CIS, 1978, 1981; Pollert, 1981; Cavendish, 1982).

The pattern of women's entry into the labour force shows that women's work is concentrated in particular local areas. This local concentration has changed the composition of the work-force and in some cases initiated new forms of political activity. Shared experiences at individual workplaces have led women to question their conditions of paid work. Struggles over unequal pay, union recognition and workplace closure have developed political awareness among women in some localities. Meanwhile similar consciousness over 'community issues' such as housing, nurseries, child care, residential homes, and more recently over the NHS and women's hospitals has linked some women's workplace demands with demands for collective responsibility for the 'domestic services' of health and education.

The experience of working women combining home and work in these localities also exposes another spatial dimension. The separation of residential areas from commercial and industrial activities, and from major shopping centres in our planned suburban areas and small towns, creates new issues for working women. Surveys in the two new towns showed that the lack of time and transport were major hurdles for working women. Rushing between home, work, shops and schools while giving attention to many different activities in one day, left women tired and with little time to themselves. The multiple tasks and the physical barriers with which women 'coped' reinforced the processes of inequality at home and at work. Many women rarely had enough time or energy to break the routine or envisage a different, more equitable, way of life.

The recession has added poverty to the worries of women. Unemployment seems to alter patterns of domestic life. There is some indication from our interviews with women in the two new towns that where both partners are in full-time employment, the burden of domestic work is more frequently shared. But when both partners are out of work, domestic labour reverts to the women. The recession seems, therefore, to be reversing some small gains.

Women's lives themselves hold the key to unlocking the complicated interrelationships between the nature of their exploitation through waged work and their oppression in patriarchal social relations. Rather than our dismissing women's own descriptions and analyses of their lives at home and at work, we should include them in our discussions. This is particularly relevant when we note that the economic life and social composition of local areas and specific towns, as well as regions, can be radically altered either by the increased participation of women in waged work or their subsequent unemployment. Both processes challenge given patterns of domestic life. Such attention to women's everyday lives is, therefore, an essential part of feminist analysis of women's employment, industrial location and regional change.

Further reading

1 For general background on women's employment and their position in the labour market we suggest C. Aldred, *Women at work*, (Pan Trade Union Studies, 1981); Chapter 2 in A. Coote and B. Campbell, *Sweet Freedom: the struggle for Women's Liberation* (Picador, London, 1982) and Chapter 7 in A. Oakley *Subject Women* (Penguin, Harmondsworth, 1981). These give clear outlines of the type of work women do. A more sophisticated analysis of why women are limited in employment choice and opportunity can be found in M. Mackintosh, 'Gender and economics: the sexual division of labour and the subordination of women' in Young, Wolkowitz and McCullagh (eds.), *Of Marriage and the Market* (1981).

2 For readings on regional inequality we propose CDP, *The costs of industrial change* (Community Development Project, London, 1977), which provides an introduction to the historical, political and socio-economic dimensions to industrial location; and D. Massey, 'In what sense a regional problem?' in *Regional Studies*, vol. 13 (1979), pp. 233–43 for her concise presentation of the varying theoretical approaches to spatial change.

3 More specifically on the geography of gender relations we recommend two general references: D. Massey, 'Industrial restructuring: production decentralisation and local uniqueness', *Regional Studies*, vol. 17, no. 2 (1983), pp. 73–91; and J. Lewis, 'Women, work and regional development', *Northern Economic Review*, Summer issue (1983).

4 If you intend doing more work for an essay or a project on this topic you will find these two case-studies useful: L. Williamson, 'Industrial restructuring, local class structure and female waged work on Merseyside', Working Paper 32, *Urban and Regional Studies* (University of Sussex, 1982); R. Hudson, 'Women and Work: a study of Washington New Town', Occasional Papers (New Series) 16, (Dept of Geography, Durham University, 1980). You could also look at *Social Trends, Employment Gazette, Labour Force Survey* and the regional publications of the 1981 Census for more specific data.

Topics for discussion

1 In what ways has gender inequality been reflected in and reproduced by the differential regional growth of women's job opportunities in post-war Britain?

2 Discuss the implications of state intervention in the economy and in family life for women's employment prospects?
3 The availability of spatially specific female labour reserves has influenced the nature and distribution of post-war industrial activity. Discuss.

5 Access to facilities

Introduction

In advanced industrial societies, the state, the market and the voluntary sector have taken over a wide range of services that used to be provided in the home. For example, education, child care, health and geriatric services are now available both as publicly funded facilities and as commodities to be bought in the private market. Consequently, the separation of home and work, private and public spheres, raises questions of access to these and other services.

In this chapter we will discuss women's access to facilities and in particular their access to public sector services. We use the health service as a case-study to illustrate how a specifically feminist analysis can be used to examine, first, what women need from a given public service; second, their difficulties in obtaining the necessary attention; and finally, possible solutions to their current problems of 'access'.

Definitions of access are usually confined to physical and economic aspects of mobility. But a brief review of the literature shows that in discussing women's access, any definition must include the limitations imposed by gender, class and race. Studies of unequal access to both private and public facilities have a long tradition in geography. Indeed, many attempts have been made to 'model' the distribution of facilities using 'optimal' location patterns based on pre-defined parameters: for example, constant transport costs, constant personal evaluation of time spent travelling, and so on. However, other non-economic concepts, such as the family, or gender, or role of facility users are not considered in this type of mathematical modelling. This methodology is therefore less than adequate in addressing the problems of women's access.

Since the early 1970s, 'welfare geography' has developed as a distinct approach within the discipline. Based on theories of welfare economics, it examines questions of *social* inequalities. One of the earliest contributors to this field was David Smith, with his concept of spatial variations in 'social well-being' (see Smith, 1973). He later published a comprehensive text designed to reinterpret the whole of geography within a 'welfare' framework

(Smith, 1977). Another example of this work is provided by Coates, Johnston and Knox (1977). But perhaps the best known and most influential book relating to questions of access to resources (indeed, one of the most significant works in relation to the historical development of geography as a whole) is *Social Justice and the City* produced by David Harvey in 1972. This study moves from a liberal analysis of the conditions for a 'just' allocation of urban services and facilities, to a more radical approach advocating a fundamental redistribution of social wealth.

This increased consideration of spatial injustice and the problems of access to facilities derived in many respects from contemporary political concerns. But despite a general awareness of sex discrimination, women's particular problems of access to and use of resources were not included. In fact there has been little reference made, within geography, to inequalities of access between men and women, or indeed between women of different classes. However, there are a few notable exceptions, and in the next section we discuss some studies which deal with certain aspects of the gender inequality of access.

Restrictions on access to facilities by women

Distance and mobility

It is widely recognized that the mobility of many households has improved very greatly since 1945, due largely to increased car ownership. Nevertheless, many geographical studies do ignore an important aspect of this increased mobility, which is that *individual* mobility is not the same thing as *household* mobility. In particular, there is a great difference between the levels of access to a car of women, men and children. Even in households with a car, it is not available to the majority of women for use during the day. Hillman (1970), who was amongst the first to recognize this discrepancy, stresses that women, and especially those caring for young children, have specific problems of mobility. Furthermore, they cannot easily substitute public for private transport. Consequently, this group of women particularly requires *local* facilities, within walking distance; both public services such as health services, play-parks, libraries and sports centres, and private facilities such as banks and shops.

However, the *actual* pattern of service provision does not reflect this need for local facilities. Instead, there is a growing tendency for services to be centralized. As service activities and administrative functions are concentrated, peripheral locations are often chosen and catchment areas enlarged. Consequently, in recent years, specifically car orientated patterns of activity have developed. For example, the growth of hypermarkets and suburban shopping centres has not only altered the location of retail activity but also the nature of shopping. Weekly or monthly 'expeditions' to buy in bulk are now common but necessitate private transport. Retailing is not the only sector affected by this relocation and reorganization; public administration (for example, tax offices, job centres) and health care, especially maternity services, have been equally transformed. We would therefore argue that this dependency on the private car has increased gender inequalities in access. There have been attempts to provide local services, through planning control, in many new towns and planned post-war suburbs. But even here, provision is still inadequate in terms of quantity and quality. In unplanned, speculatively built suburbia or on new urban peripheral estates, the situation is often far worse. Provision of local services or improvements to local mobility are, however, only partial (and often contradictory) solutions to women's problems of access. Indeed, such 'solutions' may only present further problems by reinforcing women's isolation and restrictions in dominantly residential environments (see Chapter 3, pp. 57–63).

We have outlined the inadequate location of services and the relatively poor mobility of women. However, we must now ask why the location is inadequate and why mobility is poor. Some early work on women's access sought answers to these questions in an analysis of women's activity patterns.

'Time-geography' is one approach within the discipline which has tried to emphasize the question of restrictions on activity. Hagerstrand and the Lund geographers have included the *distance*, the *time* taken to cover any distance and the *constraints* which operate on the use of time in analysing people's activities. This time-geographic perspective allows consideration of *individual* problems relating to access to facilities, and has therefore served as one very useful method of investigating women's different activity patterns (see Palm and Pred, 1974). However, many researchers consider that its powers of explanation are limited since its analysis of social constraints is weak. A modified perspective on constraints and access by women to facilities has been devised by Tivers (1982). She classifies constraints, not in time–space terms, but into categories of 'societal' and 'physical' constraints. The most important societal constraint, she argues, is the 'gender role constraint' which restricts opportunities for both women and men by pre-determining social roles. Physical constraints, such as mobility, and activity patterns are seen as subject to societal norms. Thus, Tivers' research pushes the question of *why* individual women have restricted access to facilities beyond the simple explanations of lack of mobility and lack of provision, into an analysis of underlying patterns of societal organization. For example, a study by Pickup (1983) uses the concept of gender role constraint to show how women's access to job locations is reduced by social expectations about women's family role as well as their poor physical mobility, which itself derives from their subordinate position in the family.

Cost of facilities

Many services and facilities are paid for through direct and indirect taxation, but are available free to the user at the time of use. Examples here would include urban and country parks, general practitioner (family doctor) services, hospitals and health centres, and state education facilities. But not all services are free of charge, and the cost of these may prevent or hinder access; for example, dental care and many facilities in the general sphere of 'leisure', such as sports and adult education facilities. It has been shown that, in the case of recreation and leisure activities, an increase in the amount of disposable income available is directly related to an increase in activity (Hall and Perry, 1974). In addition, the cost of travel (by either public or private transport) may limit access to desired services. It has also been shown that, on average, women are financially more restricted than men, and therefore are less able to pay for services. This has certainly influenced women's use of dental services, and increased prescription charges also threaten women's access to good health care.

One of the most important costs which many women must face if they are to enjoy reasonable access to health, education and leisure facilities, is the cost of child care. Very little support is offered by the state in this area. The majority of young children are cared for, when not at home, within the private sector – in playgroups, or by child-minders or babysitters. The lack of subsidized, high quality, state child care services, including nurseries and creches, also hinders women's choice and access to paid employment. Prohibitive costs in the private sector make low waged employment uneconomic (see Chapter 4). On the other hand, due to accessibility problems, and the constraints of young children, private child-minding is often the only source of income for house-bound women.

The high cost of excellent child care means that many women are forced to use poor quality facilities, such as playgroups without trained staff or nursery classes with high pupil:teacher ratios or child-minders who are not registered with the local council. This often results in guilt and worry on the part of women using these facilities and poor physical or emotional development on the part of the children. Needless to say the latter problems are often seized upon as reflecting the 'evil' of women's demands for an equal role in society as a whole. (Two good accounts of the problems of child care provision are Tizard, Moss and Perry, 1976; and Jackson and Jackson, 1980.)

Knowledge of opportunities

So far, we have discussed how problems of distance, mobility and cost have prevented women's equal access to health, education and leisure activities as well as to employment opportunities. However, an equally pertinent issue for an analysis of gender inequality is people's knowledge of what services exist and how to use them.

If people are ignorant of the location of facilities, whether they actually exist or not is largely irrelevant in terms of their activities. Similarly, incorrect information may also restrict access to facilities. How people *perceive* spatial opportunities determines to a large extent how they will use them.

The domestically oriented lives of many women do not provide a very good basis for accurate knowledge of what facilities are available in the outside environment. This is often particularly true in the cases of specialist mental health care, social work services and leisure services, which are considered 'peripheral' to daily life. Detailed knowledge of facilities and how to get the best out of them may also be deliberately withheld from users in the interests of 'administrative efficiency'. In other words, administrative problems in matching clients to services would be overwhelming if everyone knew in detail about every social facility and how to use it. A particular example of this is the lack of publicity about the many aspects of local authority social services or about social security benefits available from the DHSS. This lack of knowledge is further compounded by the obscure language in which official explanatory leaflets are written. Below, we discuss in some detail women's access to the health service.

Women and health: a case-study of access to facilities

A discussion of health service provision is particularly appropriate when focusing attention on women's problems of access to facilities. It is an example of public service provision which is of great relevance to women, both because of their own special needs (for obstetric and gynaecological care) and also because they tend to be largely responsible for the health of their children. Also, criticism of the male dominated nature of current health care and its implications for the control of women's bodies, has been a major current within the modern feminist movement. However, although much feminist discussion has focused on the quality of care, little attention has been paid to women's access to existing services. On the other hand, geographical work on health has examined questions of access but not, in general, *women's* access.

Current geographical research has focused on the special problems faced by people living in rural areas arising from increased centralization of facilities (see Haynes and Bentham, 1982), and on intra-urban inequalities in provision (see Knox, 1978). However, geographers initially concentrated on the state of health of territorially-defined populations, the spatial distribution and organization of health services and the access of populations to them (Smith, 1982). Many of these studies have involved sophisticated location/allocation models in order to work out the optimum location of health services and the consequences of reorganization for the access of different groups of people. However, the concept of access has usually been limited to physical mobility and distance. Even if one then adds the personal mobility of the user, this view of accessibility fails to take account of many of the problems which women encounter in their attempts to use health services.

Before we discuss whether women's access to health care is adequate, we must emphasize that the answer to this question will depend on our view of what constitutes good health. 'Health' needs to be thought of as not only the presence or absence of disease but also as the physical and mental well-being of an individual or group. The definition of health and illness varies from one society to another and therefore health should be viewed within the context of the whole society (Eyles and Wood, 1983). Hence the maintenance of good health and the prevention and cure of illness cannot simply be sought within one narrowly defined system of medical health care. Differences in people's chances of good health are rooted in the general environmental, economic and social conditions of a social group as well as in its access and ability to make use of medical facilities. Access to medical facilities is not necessarily the same as access to health.

Women have greater contact than men with the health service due to the current control and administration of contraception and childbirth by the medical professions. Women also have primary responsibility for taking babies and children to health care facilities. Their demands on the health service are therefore different from men's. However, women's experience of medical care is often unsatisfactory. For example services for 'well women', such as birth control advice and maternity care, often treat these women as if they are 'ill' and therefore 'patients', rather than as healthy women wishing to control their fertility. But, as we shall see later in this section, women have little influence on the types of health services provided, whether as members of the medical professions and health care workers or as users. It is only recently that women have begun to make more demands, through self-help and health pressure groups, for a better understanding of their health needs, and for a change of attitude on the part of the health professionals. Unfortunately, cuts in regional health authority budgets tend to undermine progressive steps to change the nature of the service, and energy is diverted into campaigns mounted to defend the services (however inadequate) which do still exist. The closure of women's hospitals, the stagnation of domiciliary services and the lack of financial and resource backing for 'community care' all have implications for women's access to *good* health care.

In Britain, the persistence of inequality in health care, despite the existence of a National Health Service based on the principle of equal access for all, has been the focus of a great deal of sociological and geographical research. Class inequality is generally measured by variations in mortality rates and different uses of health services. It is also associated with large spatial variations in expenditure and provision both between and within regions. An 'inverse care law', whereby areas with the greatest need tend to receive the fewest resources, has been postulated. Thus emphasis has been placed on inequalities of health arising from variations in general 'life chances' between classes, inequalities which are reproduced and reinforced by class-specific patterns of resource location.

There has been a disturbing absence of consideration of women's problems in this research and of policy discussions which directly address women's needs for health care. When women *are* taken into account they usually figure merely as members of a social class (their husband's) or as users of a given service, and not in terms of their own needs and treatment. Townsend and Davidson's presentation of the Black Report (Townsend and Davidson, 1980), probably the most important statement of health inequalities in Britain in the past decade, only makes passing reference to women's and men's differing mortality rates. It is also significant that in their discussion of trends of morbidity (rates of illness), they give greater details of male rates. Even though women have *higher* rates of recorded illness, illness among men is considered more important as it results in absence from paid work. Needless to say, *women's* absence from paid work is taken less seriously. The relative lack of concern with women's morbidity and how women of different classes or ethnic groups view illness prevents a fuller understanding of women's health problems in conventional studies. Fortunately there is now a growing feminist body of literature on the relationship between gender and illness (Clarke, 1983).

One important study that does focus on women's access to health is that being carried out by Vanessa Coupland. She is examining ways in which gender and class act as constraints on the access of women with young children to medical facilities. Her aims are to establish the extent to which women and young children have limited access to health services, and how far this can be attributed to gender–role constraints and to the relationship between gender and class constraints. Three hundred women, each with at least one child under five, have been interviewed in Tower Hamlets Health District (City and East London Area Health Authority) and the East Roding Health District (Redbridge and Waltham Forest Area Health Authority). Coupland found that time was a crucial element: not only the time taken to get to and from health facilities, but also the inconvenience of limited opening hours and the time spent waiting to be seen. A different study of over 100 women and their antenatal care in Islington confirmed the importance of such factors (Thornhill Neighbourhood Project, 1980). Women often had to find someone else, a neighbour or relative, to look after their children, while they attended antenatal classes. The experience also put a great deal of strain on both women and children. It appears that in many inner city areas it is not physical access that poses a problem, but rather the inconvenient opening hours and poor quality of the primary and community health services. An absence of surgeries and long journey times are, however, more important constraints on access for women living on peripheral housing estates (Knox, 1978).

It is perhaps not surprising that there is a general lack of attention to women's needs in research on health, given the male dominance of both the medical profession and of the social sciences in general. As the Women and Medicine Group have concluded, 'the majority of workers in the NHS are women, and at least 50 per cent of its potential users are women, yet the control of the service lies exclusively in the hands of men'. An examination of the American health system reveals a similar patriarchal structure.

In Britain it is in the prestigious acute sector (i.e. surgery and specialisms such as coronary care) and in the higher echelons of health administration that decision-making by women is at its lowest. Historically, the 'medicalization' of reproduction and other aspects of preventative health care, as well as the increasing use of technology, have pushed women into the background as carers rather than curers (Ehrenreich and English, 1979). In 1978, 95 per cent of the 94,000 nursing auxiliaries were women. But at the top of the medical hierarchy only 11 per cent of consultants and 23 per cent of the membership of public boards relating to health, appointed by the DHSS, were women. Over half of the public bodies listed had either only one woman member or none at all. Further gender segregation occurs among the sub divisions of the medical profession: women constitute 20 per cent of doctors in hospital service, 55 per cent in community medicine and 17 per cent in general practice. This shows a concentration of women doctors in community medicine, a branch of medicine that often has the most flexible working hours and more opportunities for part-time work than other branches. In community medicine 3229 clinical medical staff are women (59 per cent) of whom 1002 are in part-time posts. However, even in this area of medicine where women are in the majority, they are still poorly represented in the higher grades of community services. There is currently (1984) only one woman Regional Medical Officer in the fourteen Regional Health Authorities, and only two women Area Medical Officers in the ninety-five Area Health Authorities.

In hospital medicine, 7044 doctors out of a total of 35,352 are women. Gail Young in *Women, Health and Reproduction* (1981) describes the hospital subculture as 'basically a male culture'. Further she states that in order to enjoy success within this culture, women doctors have to cultivate male attributes – 'aggressiveness, competitiveness, rationality. . .' Many doctors, men as well as women, are uncomfortable in this environment but many more women doctors than men find a way out by entering other, less pressurized, branches of medicine. One reason is that postgraduate medical training is not easily combined with bringing up young children which is still primarily the responsibility of women. The areas where women might have something special to offer, for instance obstetrics, gynaecology (personal experience) and surgery (manual dexterity), are all dominated by men. In 1980 the largest percentage of women consultants was found in two specialties, anaesthetics and pediatrics, which had 20 per cent and 16 per cent of women respectively. General surgery is outstanding for having the lowest number of women consultants (0.7 per cent).

A frequently used argument against training more women doctors is that they tend to leave their career to have children and do not return to medicine. However, surveys of women doctors, which are often designed to show 'wastage' of training, have found that over 50 per cent of women are working four years after graduation. This proportion does fall between eight and twelve years after graduation. Hence the career pattern of women in medicine shows two peaks; one in their twenties and one in their forties with a trough around thirty. An early survey, undertaken by the Medical Practitioner Union (1966) found that of the 19 per cent of women who were not working, most had young children and would have liked to have worked if there had been suitable part-time work or posts with flexible working hours. There were also considerable regional variations in the percentage of women doctors with children who were not working: for example 71 per cent in Scotland, 86 per cent in Birmingham. An explanation of this variation could be linked to local differences in patriarchal relations and therefore in attitudes to working mothers (see Chapter 4 for variations in regional activity rates).

So far as the nursing profession is concerned, the ratio of men to women in the State Registered Nurse grade is 1:5, in the State Enrolled Nurse grade 1:10 and for nursing auxiliaries 1:19. However, more male nurses are found in the higher nursing grades and there is an underlying assumption in the current changes taking place in nursing administration that men will be more likely to become administrators (Salmon Report, 1972). Even in 1972 men occupied a third of all posts in the top two grades of all British hospitals. The 1974 reorganization of the NHS seems to have exacerbated this trend, because in the winter of 1973–4 male nurses were in the minority in the middle and upper echelons of the Salmon hierarchy, but three years later there was not a single woman nurse at Area or District level (Austin, 1977). Such discrimination in nursing operates not just against women but especially against black women (Hicks, 1982).

Thus, women are increasingly confined to the routine, least powerful positions within the health service. The majority of women health workers are auxiliaries and domestics who clean the wards and work in the kitchens or hospital laundries. This is hard, dirty and low paid work done on gruelling shift systems. However, most of these women are dedicated workers who see themselves as part of a team doing the necessary and potentially rewarding job of caring for others. But the divisive hierarchy has undermined their role, devalued their work and isolated them from decision-making. Strain and unrest has grown as administration has become more remote. Cuts in regional health budgets and calls for 'efficiency' have increased the workload for all health workers, but especially for those who have to maintain full care or provide adequate supplies of linen or equipment as material resources decline. The 'dedication' of health workers can no longer sustain them. Furthermore, the privatization of certain health services not only changes the ethic of 'equal access for all' on which the health service was built, but also brings into question both the quality of health care for women, who have the least financial resources, and the conditions under which women health workers will have to work (Neale, 1983).

Limiting the power of women health workers is only one result of men dominating the decision-making process within the NHS. Another result is the type of health care available to women. A vivid example is the case of abortion and its availability through the 1967 Abortion Act. The views of a small group of consultants (mainly men) have influenced whether or not a woman in a district or region can have an NHS abortion.

Considerable variation exists between regions and between districts within regions as to the availability of NHS abortions. In the West Midlands only 22 per cent of abortions are performed under the NHS, compared with 88 per cent of abortions in the Northern Regional Health Authority. It is clear that a change in the legislation relating to abortion has not eliminated the power of the male medical hierarchy to impose their views on women. Recently this power has been further extended by administrative changes at the DHSS. The wording of the form on which doctors must state the reason for terminating a pregnancy has been altered. Now there is no space for declaring social reasons. The spirit of the 1967 Act has thus been changed, not through debate in parliament which is at least open to pressure from national groups, but by senior (male) Whitehall civil servants.

The treatment of depression in middle-aged women is another example of male bias in the medical profession. Doctors tend to apply a double standard of mental health. Male doctors frequently have some sympathy and understanding for the problems faced by middle-aged men, probably because they can relate these to their own feelings and experiences, but they often have little awareness of what it feels like for a woman whose family has left home and so deprived her of one of her roles or her only role in life. The problems of middle-aged men tend to be seen as a consequence of a competitive and status-conscious society but those of women are more often attributed to personal inadequacies. Doctors frequently try to smooth away the worries of middle-aged women by prescribing psychogenic drugs that change moods but do nothing to alter the structural limitations of the woman's life which led to the depression in the first place (Barrett and Roberts, 1978). Similarly, the attitude and the treatment of the growing population of elderly women leaves

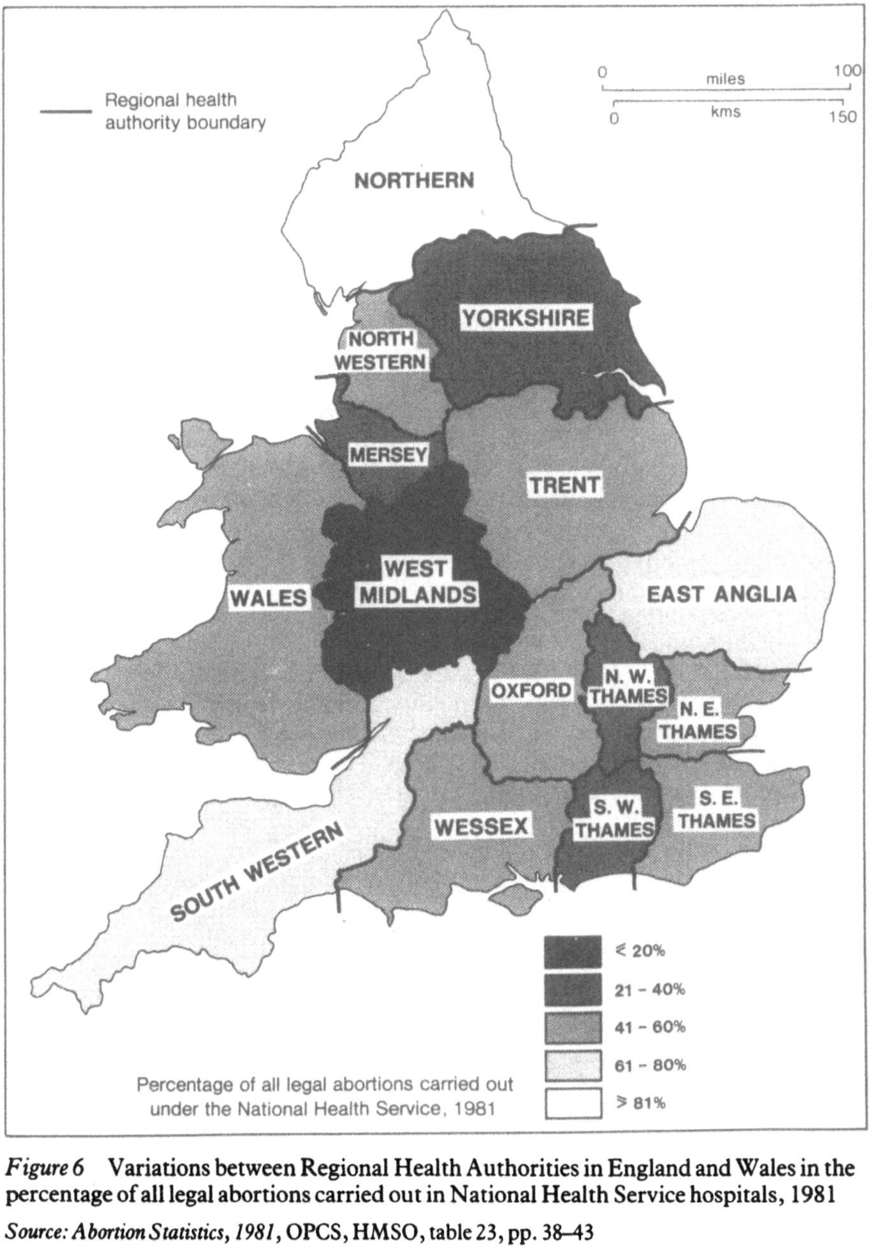

Figure 6 Variations between Regional Health Authorities in England and Wales in the percentage of all legal abortions carried out in National Health Service hospitals, 1981

Source: Abortion Statistics, 1981, OPCS, HMSO, table 23, pp. 38–43

much to be desired. Geriatric medicine has low status anyway, but the specific social, economic and mental problems of this group, who are often coping alone for the first time, are frequently underrated. Little dignity is left to these women who have replaced dependence on their husbands by dependence on the state and the NHS.

Another example of the impact of male decision-making on women is the nature of antenatal care and childbirth. In theory women should have a choice ranging from home birth through shared general practitioner/hospital care to total hospital care. However, in practice, the decision about the form of care is taken by the hospital or doctor, and indirectly by the health authority through decisions over the structure of maternity services. From 1963 to 1977 the number of home births and births in private nursing homes declined from 270,000 (one in three) to 12,000 (one in fifty). This decline was accompanied by the dismantling of the domiciliary midwife service and the erosion of the autonomy of midwives. Midwives, as independent practitioners, are legally entitled to deliver a woman on their own but in practice their decreasing influence has further limited women's control of childbirth and immediate post-natal care.

Although home delivery is still retained as a matter of policy, there is considerable spatial variation in its availability. In addition, those who decide on this option often do so in the face of stiff opposition. None of the ninety-eight Area Health Authorities exceeded 10 per cent home deliveries for the period 1974–77 (Ashford, 1981). There are many opposing views among feminists and others about the desirability of home delivery, with arguments against home deliveries usually based on claims that it is far safer for women and their babies to be in hospital. At present, there is still insufficient study of the variations in perinatal mortality (infant mortality from twenty-eight weeks' pregnancy to first week after birth) in relation to place of confinement, controlling for the quality of antenatal and post-natal care. One specific study appeared to show no differences in the levels of perinatal mortality between home deliveries, general practitioners' and central hospital maternity units (Ferster and Pethybridge, 1974). As in other areas of health care, however, it seems probable that broader economic, social and environmental conditions are the most significant determinants of perinatal mortality. From the early 1960s to the late 1970s the gap in perinatal mortality rates between urban and more rural areas, between the North-West and South-East and between social classes has not declined (Ashford, 1981, p. 102). The desire for home delivery, expressed by many women, is an index of the poor quality of personal care and treatment in the male-dominated, authoritarian environment still common in many hospitals.

These examples of male-dominated decision-making in areas of direct relevance to women as users of services, illustrate the fact that *access* to facilities encompasses more than spatial pattern. In some cases women are deliberately misled in the information they receive, or are otherwise restricted in their effective access to health care. What is currently lacking in many geographical studies of access is an understanding of the impact of gender relations on availability – that is, how the opportunities for obtaining and using a service are limited by an unwillingness to recognize either women's needs or women's ability to make decisions about their own bodies and health care, and by the constraints that family responsibilities and lack of money place on women's mobility. Better, two-way diffusion of information, the provision of support facilities, such as creches, more flexible opening hours and door to door transport, and a wider choice in terms of the form of services for women would improve their access to necessary health care.

The ways in which gender operates with class and race to restrict women to this narrow range of services, which do not take into account their needs and preferences, require much more detailed work. Although it is possible to construct a scale of the severity of constraints, as Vanessa Coupland has done for married women with young children, this can only serve as a guideline to indicate the number of constraints faced by particular groups of women. A critical feminist approach must encompass the response of women to these constraints and must therefore focus on the reality of women's health experience in relation to their gender role and class position within the family, the community and at work. We would argue that large quantitative surveys do little to examine these issues. Time spent on case-studies and talking to women about their expectations will do more to challenge social stereotypes about different groups of women. Such an approach is even more urgently needed for women belonging to ethnic minorities (Donovan, 1983). We are not, however, suggesting that we should dispense altogether with studies of physical accessibility and quantitative techniques, but that their limitations should be clearly recognized.

Lack of knowledge about women's particular needs and their problems of access to appropriate health care is illustrated in recent health planning documents. Since the early 1970s the NHS has been co-ordinating its diverse services territorially and redistributing resources among the regions. However, this form of health care planning has relied to a large extent on calculations which have been based on inadequate data and questionable surrogates of need. The Resource Allocation Working Party (RAWP), established in 1975, arose from a concern about inequalities of expenditure between the regions. The formula for the reallocation of resources was based on population, taking into account age and sex structure, and standardized mortality ratios. But the use of mortality ratios as surrogates for illness and need for health care is dubious. Morbidity data (that is, data relating to illness rather than death) would be preferable, but adequate data sets do not exist.

The consequence of this reallocation has been to divert funds from the south-east of England to the north and west, and from inner city areas to the suburbs and outer metropolitan areas. Yet although inner cities may have previously received higher levels of funding, much of this went to the acute services which have been concentrated in urban areas for historic reasons. In 1980 the London Health Planning Consortium directed its attention to the provision of acute hospital services and recommended that a reduction in inner London should be compensated for by a growth of services in the Thames counties. It is assumed that a reduction in hospital beds will release funds for other services but 'nowhere in the discussion document is the mechanism of rate of change in favour of these services discussed' (Eyles, Smith and Woods, 1982). Likewise in RAWP there is no mention of the expenditure and interaction between different services.

The implications for women of this use in health care planning of simplistic formulae and lack of understanding of the interaction between different sectors of health care are immense. First, mortality does not, in advanced industrial societies, reflect the level of health of a community nor the changing demands made upon its health services. Although women's and children's greater need for and use of health services may not be fully translated into indices of morbidity, at least the use of these indices would be an advance on using mortality data. Secondly, the recent policy supporting the substitution of community care for institutional care raises many issues concerning women as clients and workers in health care, and about the relatonships between specialized services. Caring in the community is most likely to fall on women. Professional nursing of the sick and handicapped is replaced by isolated mothers, wives and sisters who receive minimal state and limited charitable support. Job loss for low paid health workers and confused liaison between separate services, such as physiotherapy or occupational therapy, mental health services, geriatric services and peripatetic child care services, are likely to accompany this change. We have to examine the nature of the community care proposed and its consequences for women. Finally, the shift of resources away from inner cities, and the dominance of health care resources by 'acute' hospitals and their staff means that there is further pressure on primary and community health care. These sectors will have to rely more and more on inner city partnership projects, where the limited resources are allocated by central government, local authorities and charities only for specific short-term projects.

There have been numerous attempts by women to win control over their lives and bodies and hence free themselves from the dominance of male decision-makers. Perhaps the most notable in recent years has been the women's health movement, an integral part of the women's movement. An essential component of this health movement has been women's developing knowledge about their bodies and their subsequent criticism of many areas of conventional medical practice. Alternative methods have been explored in self-help groups. However, pressure groups fighting for improvements in existing services have also evolved: for example, for more free abortion facilities, for safer forms of contraception and for better cancer screening services. As the recent publicity about the rights of women in childbirth has shown, the medical profession is not completely impervious to criticism but it responds more favourably to demands made by articulate, white middle-class women. Nevertheless, attention to women's demands is still scant, and women are resorting to self-help groups for support over care of themselves and their handicapped or ill dependants.

Conclusion

In this chapter we have shown the lack of research within geography directed at the problems women face in gaining access to services and facilities. Lack of awareness of gender issues has meant that researchers have failed to ask whether the location, type and quality of facilities provided reflect or reinforce patriarchal relations. Likewise, they have failed to examine how systematic differences between women's and men's mobility, time constraints and information are related to differences in their economic and social power. Thus the orthodox approach of geographers to problems of accessibility, which takes types of facility and the mobility of different groups as given, offers only an incomplete analysis. We have discussed how gender relations affect the particular way in which women's access is limited by distance and mobility constraints, by the cost of facilities or travel, by incomplete knowledge of opportunities, or by restrictive assumptions about gender roles which are built into service provision.

The case-study of women's access to health facilities clearly shows how access may be limited by these constraints. Cost of services is not a major issue here, although there are cost implications in prescription and dental charges, in the reduction of local services, and in the need for substitute child care during medical treatment. Distance and mobility constraints are important, particularly in association with the constraints imposed by the timing of women's domestic and waged work responsibilities, as are restrictions imposed by lack of knowledge (some of which are directly the result of male decision-making and male dominance of the medical profession).

Geographers, in recent years, have been active in considering social and class inequalities, including those relating to access to facilities. But the special problems of women have been largely overlooked. It is not sufficient to consider patterns of inequality simply on a family, household or class basis. Patriarchal relations operate within and across family and class structures.

Therefore it is essential that a feminist approach to, and explanation of, access considers gender-based inequalities, both alone and in association with class and geographical factors.

Further reading

1 Preliminary ideas and a framework for a feminist perspective on the health status and needs of women with severe constraints can be found in V. Coupland, 'Gender, class and space as accessibility constraints for women with young children' in Health Research Group, *Contemporary Perspectives on Health and Health Care* (Department of Geography, Queen Mary College, London 1982).
A useful explanation of health planning reorganization and the role of the medical profession and bureaucracy can be found in J. Eyles, D. Smith and K. Woods, 'Spatial resource allocation and state practice: the case of health service planning in London', *Regional Studies*, vol. 16 (1982), pp. 239–53. This is a welcome step towards a political economy of health in geography. A good survey of the issues concerning women and health in medicine can be found in J. Leeson and J. Gray, *Women in Medicine*, Tavistock Publications, London, 1978).

2 N. Thrift, *An Introduction to Time-Geography* (University of East Anglia, Geo Abstracts Ltd, CATMOG no. 13, 1976). As the title suggests, an introduction to the concepts and applications of time-geography. The concept of a 'gender role constraint' is discussed in J. Tivers, 'Constraints on spatial activity patterns – women with young children', Occasional Paper in Geography, 6 (Kings College, Dept of Geography, London University, 1976).

Topics for discussion

1 What factors are likely to account for the variations from one area to another in the proportion of abortions carried out by the National Health Service in England and Wales?

2 Few women are in positions of power in the medical hierarchy. What implications does this male dominance have for women's access to health care?

3 What measures could be taken by the state to improve women's access to facilities? Explain which measures you think would be most effective.

Organizations to contact for information on issues relating to women and health

Community Health Council

There is one for each District Health Authority. The main activities involve linking members of the council, the public and the NHS; it protects consumer standards, provides information and advice. The twenty-four members comprise twelve Local Authority nominees, six from voluntary organizations and six from Regional Health Authorities.

Kings' Fund Centre,
28 Albert Street, London NW1
This is a foundation specializing in health projects and research. It has an excellent library that receives 200 health periodicals from Britain and abroad.

Medical Practitioners' Union

It is linked with ASTMS (Association of Scientific Technical and Managerial Staff, 10 Jamestown Road, London NW1). Doctors joining a trade union usually belong to this one. It has produced a booklet bringing together a series of articles on women's issues published in *Medical World* (the MPU Journal).

Women's Health Information Centre
12 Ufton Road, London N1

Journals

Medicine in Society carries a large number of articles on feminist issues in health care.

Social Science and Medicine is an international journal, divided into four sections, one of which is on the sociology of medicine, another on medical geography.

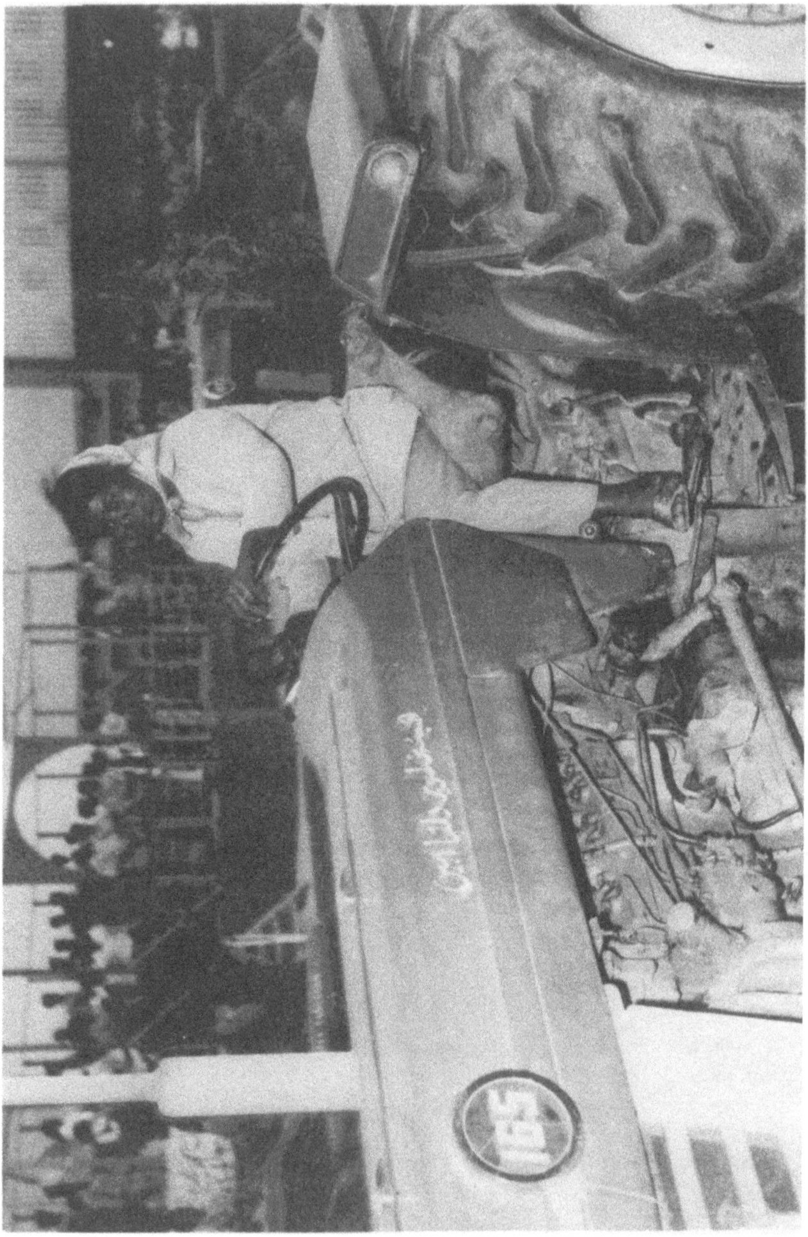

Figure 7 Even in Arab countries women's productive roles are rapidly changing

6 Women and development

Introduction

Observation of recent political and social changes in Africa, Latin America, Asia and the Middle East reveals a common thread. Women have been among the guerrilla fighters and charismatic leaders of nationalist movements; women have led protest movements against injustice; women's grievances have been reflected in the various revolutionary manifestos – yet the feminist movement has rarely been maintained (Randall, 1981). New regimes often praise the contributions of women but are slow to introduce equal rights reforms except where they are necessary to the economic goals of the government.

Most theories of development have little to say about women (Papanek, 1978). Presumably, generalized constructs such as 'the peasant' refer to both men and women, but empirical studies tend to see only men in the fields. Discussions about the impact of development policies on women see them as objects, rather than agents, of change. Whether change occurs as a result of internal revolution, technical and financial aid from an international agency, or technology transfer by a multinational company, the effect on women is rarely considered. Until quite recently, it was generally believed that economic growth and development was a sufficient condition for an improvement in women's social position. Such an approach, which is derived from Myrdal's 'trickle-down' theory of development, denies the unequal power relationships which exist between people of different castes, races and classes and between men and women.

Despite increasing media coverage of women activists in the Third World, women have been virtually invisible in the geographical literature. Textbooks provide a male-centred view of the world in which men are the producers and organizers of societies while women play a purely reproductive and supportive role. As previous chapters have shown, this view bears little relationship to the reality of life in the West today. How much less valid is this Western viewpoint for Third World countries. In this chapter we will show how women are engaged actively in changing the face of the developing world. We will also discuss how the use of Western stereotypes of gender roles has obscured an understanding of women's contribution in both our explanations of the development process and our assessment of existing needs.

Such perceptual invisibility of women is reinforced by a lack of appropriate statistics. This problem was first tackled by Boserup in her seminal book *Women's Role in Economic Development* published in 1970. Boserup was forced to use United Nations data which were widely recognized as inadequate, especially with regard to gender role differences. However, her experience in the developing world allowed her to reinterpret the statistics and to show that not only did women play a hitherto unrecognized major role in Third World economies, but also that regional differences in female roles were very marked. For geographers, this latter point is probably her most important contribution. Boserup's analysis suggests that differences in the division of labour between women and men are not based on particular cultural perceptions of what is correct or proper in relation to physiology or biological determinants. Rather she argues that social relations, as reflected in changing land tenure systems and composition of the labour force, are of prime importance in determining the gender division of labour. For example, where there are few waged agricultural workers, as in Africa and some parts of South-East Asia, women make up a high proportion of a family labour force engaged in agricultural production. In areas where wage labour is important, as in Arab and Latin American countries, women are less likely to participate in the agricultural labour force. However Boserup identifies the Caribbean as an anomalous sub-region. One would expect to find the Arab/Latin American pattern, but in the Caribbean the decision-making role of women in agriculture is characterized by an African pattern. She ascribes the importance of women in Caribbean agriculture to the African origin of the majority of the population (Boserup, 1970) although it seems probable that the observed differences reflect distinct colonial heritages rather than intrinsic ethnic variation. Again, we can agree that it is the social relations which define gender roles rather than biological sex which underlie the role of women in development.

Boserup's comparative analysis was based on observed regional differences in the composition of the labour force in agriculture. She then extended this analysis to examine patterns of women's participation in non-agricultural activities. She drew attention to the influences of farming systems on migration patterns, then showed how a gender inbalance in the rural to urban movement of peoples affected the participation of men and women in urban labour markets. These generalized links between the form of agricultural production, rates of migration and the nature of urban economic activity are at times overstated. However, this has inspired a great deal of subsequent empirical and theoretical work.

During the 1970s most of the research on women's role in developing countries was directed towards data collection and the search for empirical evidence for Boserup's contentions. This led to the consideration of a number of conceptual and measurement issues. For example, how do we define and measure the concept of 'work' with reference to women? What kind of statistics are relevant to women's participation in development? One does not have to look far to find inadequacies in the record. As recently as 1978, material presented by the United States Department of Labour in conjunction with the United States Foreign Assistance legislation suggested that only 5 per cent of African women were in the 'work force'. Yet African women, as shown in the empirical evidence cited in Boserup, do 60 to 80 per cent of the work in the fields and at times put in 16-hour days. Official surveys define 'work' as waged employment. They therefore exclude the unpaid subsistence and domestic labour which women do within the family or local community. The published national and international statistics which are specifically concerned with women tend to focus heavily on their role in biological reproduction, rather than on their economic or social participation. Nevertheless, a recent world fertility survey showed that at the time of data collection over 70 per cent of women under 50 years of age were neither pregnant nor lactating. One-third of women are therefore involved in biological reproduction in any one period. The consequent drain on their strength, energy and time is bound to affect their specific productive and economic role. However, the assumption that *all* women are solely engaged in biological reproduction has implications for every woman's social and economic role in the developing world. This link between women's activities in production and reproduction has, however, been largely ignored in the literature.

Another aspect of a Western male-centred view is reflected in the statistics collected in and on Third World countries. The term 'head of household' is usually considered to refer only to men, although in many Third World countries over one-third of the households have *de facto* female heads (Buvinic and Youssef, 1978). The common census definition of work also belittles women's contributions. Work of less than twenty hours a week or seasonal work inputs are often ignored. Furthermore Rogers (1980) suggests that women's work is valued at only two-thirds of the time actually spent while men's work is fully counted. Patriarchical assumptions about the sexual division of labour also affect the classification systems used by census takers and thus contribute to an under-reporting of women's activities. In addition, the focus by national and international statisticians on occupations in the formal cash economy contributes to an underestimation of women's work because of the preponderance of women in informal activities.

By 1980, when the United Nations Conference on Women was held in Copenhagen, some of the early myths about women's role in development had been overthrown. At this conference it was accepted that the female half of the world's population provides one-third of the official labour force but carries the burden of two-thirds of the hours worked. Women throughout the world work in agriculture and provide 44 per cent of the world's food supply. For this contribution women receive 10 per cent of world income and own a mere 1 per cent of world property.

The impact of development on women

Boserup feels that women are victims of development. She suggests that this victimization occurs for three main reasons. Firstly, because more women than men engage in those traditional activities (such as pottery and weaving) which are most likely to disappear with modernization. Secondly, women suffer since they are less free to adapt to new conditions because they usually have to face discrimination within the labour market. Finally, women generally have to cope with family obligations which make them less mobile than men. As we have seen in Chapter 4, these patriarchal attitudes which restrict women's productive roles are common to most contemporary societies. In addition, the improved access of children to education makes them less available to their mothers for assistance in household chores.

Modernization trivializes the role of women by bringing about changes in cultivation systems and forms of property holding as well as by the introduction of new crops and technology. It was the decline of land as the primary source of capital and its replacement by industrial capital in the post-feudal era that eroded the power base of European women, leaving them dependent and politically helpless by the nineteenth century. Yet the significance of the loss of land and its prerequisites as a source of capital for women, a process now occurring in industrializing Third World countries, has been missed. Land reform is often detrimental to women as it tends to reinforce the patriarchal views of officialdom. In agricultural co-operatives in North-East Brazil only sons, not daughters, may inherit the family holding. On modern irrigation colonies in the same region women, widowed or abandoned by their husbands, are forced to leave the colony unless they have a son old enough to take over as official head of household and colonist.

Where agricultural modernization takes place there is an increased emphasis on cash crops, usually grown by men, with a concomitant decrease in women's subsistence production as land and labour resources come under pressure. This change in the land use pattern often leads to a decline in family nutrition levels and in the women's independent economic base. Such a situation has been well documented by Stavrakis and Marshall (1978) working in Northern Belise where the introduction of commercial sugar production brought sudden changes in the family economy. Men abandoned maize cultivation and hunting in favour of sugar cane growing. The cash income they spent largely in the local rum shops. Rubbo (1974) showed how, in Colombia's Cauca Valley, the introduction of maize, soybeans and tomatoes, using high levels of modern inputs, to replace the traditional coffee and cocoa trees, reduced the land available for women's subsistence production. Women thus became increasingly dependent on men's access to wage labour.

The intrusion of cash-cropping and export agriculture into pre-industrial economies is always accompanied by an upheaval in both gender and class relations. Young (1978) demonstrates, with the aid of a Mexican case-study, how women's productive work is affected by changes in the relations of production; while Stoler (1977) drawing on experience in Java, and Deere (1982) using Peruvian data, show how women's productive role varies according to their class position. Whereas much of the empirical work has convincingly demonstrated that women's socio-economic position does not necessarily improve with economic development, the supporting theoretical work, critical of women's oppression and exploitation under capitalism, has idealized women's roles and socio-economic position in pre-capitalist modes of production. Anthropologists who follow Margaret Mead have taken an evolutionary approach suggesting that women's status was high in most traditional societies. The structure and ideology of male dominance they claim was an artifact of colonialism. Recent criticism and evaluations of Mead's work suggest that such a 'noble savage' approach is an over-simplification in which the observer's own cultural bias has led to erroneous conclusions. Traditional societies hold a great variety of attitudes to relationships between men and women but the world-wide impact of European culture did lead to the spread of patriarchal beliefs. Engels associated the acquisition of private property with the origins of women's subjugation. Modern extensions of Engels' thesis tend to stress the development of the social relations inherent in capitalism as the single most important factor in explaining women's subordinate position in relation to men. Deere (1981) argues that women's economic participation in the Third World differs significantly from their economic participation within the centre of the world capitalist system. In the periphery the majority of women live in rural areas and are active in subsistence farming as well as in petty commodity production and trade. Women's family maintenance role encourages the payment of male wages which are insufficient for the reproduction of labour. Even in urban areas female involvement in the informal sector, and in such activities as firewood gathering and peri-urban agriculture, enables them to continue their economic support of the family. Thus the separation between home and work, between private and public spheres of activity, which has characterized gender roles in the industrialized world since the Industrial Revolution is much less clear in the Third World. The world-wide effects of capitalist development on women are seen most clearly when contrasted with the situation in state-planned 'socialist' economies as in Croll's (1979) study of the changes that have affected peasant women in the Soviet Union, China, Cuba and Tanzania. These economies have encouraged women's paid employment and have provided education for women, the means to control fertility and child care facilities. However, because these state-planned societies have failed to ease the burden of household chores, women have tended to resist incorporation into the wage economy. Although the socialization of housework has not been achieved, women's public status has improved. The long-term *aims* of these state-planned economies generally appear to be in women's interest but in the short term state-planned development has tended to increase women's burdens by expecting them to fulfil a dual role in both public and private worlds and by introducing state control of their biological role. Empirical evidence suggests that no single approach to development is clearly beneficial to women. In state-planned Tanzania the villagization scheme reduced women's access to fuelwood supplies while still leaving them responsible for this aspect of household maintenance. In right-wing Bolivia, Aymara women took advantage of the commercialization of agriculture to gain economic strength and move out of their traditional subservient role. It appears that local leadership and co-operation between women are two of the most important factors determining regional variation in the impact of development on women.

The gender division of labour in agriculture

In hunting and gathering societies gender roles are clearly defined and women provide 60 to 80 per cent of the daily food intake. Crop cultivation and the domestication of animals add new complexities to the gender division of labour. In these horticultural societies men clear the land and women do the planting, weeding and harvesting. Women generally raise small animals, such as pigs and chickens, which can be kept close to the home and fed on scraps. This activity is regarded as an extension of their household duties and often provides personal savings and income for the women. Men and boys assume control of larger animals such as cattle, sheep and goats which need to be pastured at some distance from the home. The invention of the plough changed the methods used in crop cultivation and led to the redistribution of the related tasks, with a greater proportion of farm jobs being subsumed by men. However, even today, on many small farms in the Third World, the major tools are still the hoe and the cutlass and there is little evidence that changing technology has influenced the division of labour. On larger farms where commercial crops such as sugar and rice are grown, the rapid spread of mechanization has affected the traditional tasks undertaken by all members of the farm family (Bagchi, 1982). The recent introduction of modern inputs such as pesticides, herbicides and chemical fertilizer has also influenced gender roles. Women very soon became aware that these chemicals were dangerous to their health and their reproductive success, so that even on the poorest farms operated by women, a man will be employed to apply pesticides (Harry, 1980).

In addition to these evolutionary changes which have occurred throughout the world, political developments have brought new forces to bear on gender roles in many Third World countries. Colonialism and the plantation system introduced major modifications. The strict regime of long hours of supervised labour in the fields, factories and estate houses made equally heavy demands on both sexes. Yet the very rigidity of the system, dependent on external control and demand, led to the development of internal variations. The cost of feeding a large slave labour force persuaded the planters to allow peasant-like activities to develop. This growth of marginal production and internal trade within the plantation slave economy, with its concomitant gender division of labour, occurred to varying degrees in many areas dominated by this form of agricultural organization.

In post-slavery plantation economies, women found it increasingly difficult to obtain paid employment on estates. For example, in Jamaica the steady decline of the agricultural labour force, especially after 1890, forced rural women to migrate to the towns to seek employment in the service industries. In other parts of the Caribbean the decline in the agricultural labour force did not occur until after the Second World War. This change in the sectoral distribution of employment was accompanied by changes in the occupational pattern of women agricultural workers. In 1946 women made up 42 per cent of the unpaid workers on West Indian farms and 47 per cent of the paid agricultural workers. However, by 1961 these relative positions had been reversed, with female paid employment declining and the proportion of women in the unremunerated category having risen to 59 per cent. Garrett (1976) noted a similar change in Chile over the same period and suggested that this represented a serious deterioration in the ability of women to obtain those sorts of agricultural jobs which would allow them to support themselves and their children. In the last decade the most striking development in the West Indian labour force, as in many other parts of the Third World, has been the rapid increase in the female participation rate. This has not, however, occurred in the agricultural sector. Between 1961 and 1971 the proportion of women working in agriculture fell by approximately 5 per cent in Barbados and St Vincent and by as much as 25 per cent in Antigua as alternative employment became available. Many women moved into white-collar jobs, with 28 per cent of the working women in Barbados, Trinidad, Guyana and Jamaica in this sector as compared to only 13 per cent of the men. Women in the Caribbean have also been able to take advantage of the new employmnt opportunities provided by development projects in tourism and manufacturing. Monk (1981) has shown how, in the coffee-growing regions of Puerto Rico, government programmes encouraging the dispersal of industry into rural areas have provided jobs for women rather than men. Nevertheless, the current economic recession appears to be hurting female non-agricultural employment more than that of men.

In state-planned economies it has become public policy to attempt to reverse the decline in female participation in agriculture. For example, prior to 1959 only 2 per cent of the Cuban agricultural labour force was female but, because of a post-revolutionary shortage of labour in rural areas, women have been encouraged to become paid agricultural workers. However, as a result of the seasonal pattern of demand associated with the major Cuban crops, they have tended to form a reserve labour force in both collective and private sectors. In the 1960s, 55,000 peasant women received training in agricultural skills and in the following decade twice as many women were taught to raise livestock and vegetables for sale to state agencies. Although the Family Code of 1975 formally enforced the principle of the sharing of domestic tasks when both marital partners were employed, there is no evidence that this has yet occurred. In general, Cuban peasant women have resolved the conflict between their domestic duties and the demands of the farm by co-operating among themselves in matters of child care and refusing to take on other paid employment outside the home. When Jamaica and Grenada were under their socialist premiers (Manley and Bishop respectively), and were faced with food shortages, underutilized land and unemployment, it became government policy to give women leadership roles in agricultural production and marketing.

It is not only the type of agriculture or the political complexion of the government which can influence the role of women. Even more important in multi-racial societies may be the relative status of women within the various ethnic groups and the cultural-historical background of individual communities. Above all, the day by day occupations of women are determined by economic pressures and the impact of migration on sex-ratios in the population of the country as a whole and within different regions.

Women and migration

One of the most pervasive trends in Third World countries today is that of migration from rural areas to cities. However, this migration does not affect both sexes equally. In Africa and the Far East male migration is dominant but in Latin America women are more likely to move than men. This movement of people is a response to changing spatial patterns of growth, both real and perceived, as industrialization occurs in Third World countries. It has far-reaching effects on agriculture. In rural areas where women and children are left behind to cultivate the family farm, land use may change from commercial to subsistence crops (Momsen, 1981) and the introduction of modern technology may be unsuccessful where women are not included in training programmes.

Rengert (1981) has shown that in Mexico as a whole only eighty-three men migrate for every 100 migratory females. This difference may be explained in terms of push and pull factors. It is easier and less costly for women to migrate to the city as many moves are related to marriage or domestic employment. If they move with their husbands then the husband bears the cost. If they move to a position in domestic service, then transport, lodging and sustenance are usually provided by the employer and the migrant may be able to contribute to the family income. In the case of men migrants, the rural family not only loses a worker but also has to bear the cost of migration – although the out-migration of male children often results in more income for the individual than would be possible in the village. The higher rates of migration of sons from smaller or wealthier families suggests that migration of male children may be seen as an investment aimed at the socio-economic advancement of the whole family. Where women are able to make an economic contribution to their families while remaining in rural areas, they do not migrate. This is the situation, for example, in the Mexican village of Tzintsuntzan, where the women have traditionally been highly skilled potters.

Differential migration leads to an increase in female-headed households and affects gender roles. Sando (1981) in her work in Taiwan, found that depopulation of rural villages affected men and women to different degrees. Women were drawn into more farm work and began to undertake a wider variety of tasks. Even where the husband was present, if farm work was performed by only one spouse it was most likely to be done by the wife. If the husband had a job off the farm but continued to reside within the village, he was considered officially to be a part-time farmer although all the farm tasks were commonly left to the wife. Women were also more likely than men to work for a wage on other people's land.

In the cities, women are more constrained than men in their choice of occupations. Most women go into domestic service or petty commodity production. As domestic servants, they may free upper-class women for professional employment but in their turn find it difficult to care for their own children. If they work within their own homes providing services or home-produced goods, they are spared the expense of hiring babysitters. Alternatively, women may take their older daughters out of school in order to care for younger siblings and free their mother for waged employment. Nelson (1979) in her study of a squatter settlement fifteen minutes drive from the Nairobi Hilton in Kenya, found that most women worked from their own homes, either offering sex or illegal home-brewed beer for sale. Women's occupational choices are more restricted than those of men. They commonly have less education, fewer marketable skills and so have less access to capital. African women, for example, are seen primarily as mothers and generally are spatially constrained by the presence of young children. The most successful women are either those who are barren or those who have adult offspring who are able to assist them financially. Women in the informal economy of Third World cities are usually poor and overworked, but they are possibly less exploited than they would be if they sought a position at the lower end of the formal economy or sought a partner with work in the formal economy. Female urban migrants suffer from the loss of support of their rural extended family but the city may be the only option available for women who are widowed, divorced without hope of remarriage, or have children out of wedlock.

Women in industry

Industrialization in Third World countries is opening up new opportunities for women but its impact is not always beneficial. As manufacturing has changed from handcrafts based on the family as a production unit, to factory based industries, women have found it more difficult to participate at the higher skilled levels. Much of the new industry, however, is labour intensive processing and assembly of goods for export, often controlled by overseas capital. This new international division of labour is having a major impact on the gender division of labour in Third World countries.

The limited data available suggest that the effect of these foreign-based assembly industries on the division of labour in Third World countries has mainly been seen in its exploitation of a new category of workers; that is, young unmarried women who form the majority of the workers and are employed as unskilled or semi-skilled labourers. The effect of this employment on traditional family structure appears to vary with the degree of male unemployment and the strength of the patriarchal tradition. Studies of the Mexican Border Industrialization Program reveal that the majority of men living in households of female assembly operators are unemployed. The Program was set up to provide jobs for men dislodged by the ending of the *bracero* programme. The *bracero* programme was set up by the United States government in the early 1940s when the war effort caused labour shortages in the United States, especially in agriculture. Mexican men were recruited on short term contracts, and sent to work north of the border. The Border Industrialization Program encouraged the location of assembly plants in Mexican towns located close to the border with the United States. By 1979 there were 540 plants in operation employing 110,000 workers and contributing $759 million to the Mexican balance of payments. However, the Program uses only 1.5 per cent Mexican raw materials in its industrial activities and about half the workers' incomes is spent in cities on the US side of the border. The Border Program has not only failed to provide jobs for former *braceros* but has also changed the whole structure of employment in the region. Christopherson (1980) shows that about 80 per cent of the labour force employed in assembly operations in the border town of Juarez is composed of single women between the ages of 17 and 25. These young women often earn more than their fathers, or they may be the only breadwinner in the family. Under these conditions the families put great pressure on their daughters to make sure that they perform properly at work and so do not lose their jobs. Such family control, which ensures the docility of the workforce, is of great benefit to the foreign firms operating in the Border Program. But this employment pattern seems to have had a disruptive effect on traditional family life. One-third of the women employed as garment manufacturers in Juarez are now heads of households (Safa, 1981). These women have achieved some 'independence' but this is limited by their poor wages and working conditions. By contrast, in Puerto Rico, where there is also a strong Hispanic patriarchal tradition, Safa has failed to find any correlation between female employment and marital breakdown. The absence of this link is less surprising in Jamaica, where a tradition of female-headed households dates back to the days of slavery and the majority of female factory workers has no resident spouse. In the Commonwealth islands of the Caribbean, among the Afro-Caribbean population, there is a long history of women working outside the home and being economically self-supporting, so the new industrial employment has been less challenging to existing family life than in many other areas.

Export-orientated agribusiness, such as the cashew industry of North-East Brazil or strawberry growing in Northern Mexico, also depends on women for much of the gathering, processing and preparation of the products. The local men have often migrated to better paying jobs elsewhere. This type of work allows women to remain in rural areas and thus supports the spatial stability of the family. In other parts of the world, such as Malaysia, families remain in the countryside while the young women migrate alone to live in the city, in single-sex boarding houses, and work in the factories.

The impact of this new export-orientated industrialization on the gender division of labour also varies with the age and marital status of women workers. Most women recruited to work in modern industry are young and single. They contribute to their parents' income rather than maintaining a household of their own. This pattern tends to be less socially disruptive than employing married women, in that it does not usually challenge directly the role of the male head of the household – although, as in the case of Juarez, it can do so. Thus, the 'emancipation' of these young women, gained through waged work, is limited by their family obligations as it is also by the ill-health which often accompanies this form of employment (Phillips and Taylor, 1980). Older women working to support their children are more likely to be dissatisfied with low wages and so many companies discriminate against them. This discrimination is often reinforced by national governments who feel they need to keep the level of wages low in order to prevent these footloose industries moving elsewhere to countries with an even cheaper supply of labour. Control takes the form of the prohibition of strikes as in Brazil, or the provision of income subsidies such as the food stamps offered in Puerto Rico. On the whole, throughout the Third World the underpaid employment of female members of the family is used to cushion the effect of male unemployment resulting from changes in national economic structure brought about by development.

New directions

Geographers have only recently become interested in the problems of women and development. The first book by geographers on the topic was published in 1981 (Horst). The International Geographical Union had its first seminars on this topic in 1981 and 1982 and the Institute of British Geographers discussed it for the first time in 1984.

As geographers take a fresh look at the developing world, some surprising grassroots developments among women themselves are revealed which may yet undermine the logic of Boserup's original thesis. Jane Benton's restudy of Aymara women in Bolivia's Lake Titicaca region showed how the position of women had been transformed between 1971 and 1981 (Benton, 1984). Marriages are no longer arranged with land and livestock prospects uppermost in the mind. A number of girls express a wish for children for company but are adamant in refusing marriage. They do not intend to place themselves in a situation where, like their mothers, they could be regularly beaten by drunken husbands. Male attitudes to secondary education for daughters have changed since it became obvious that educated girls could gain higher status employment and increase the family income. Many of the successful female weekend onion vendors of 1971, a decade later were operating lucrative small businesses employing men. Women are now using machinery formerly denied them on the grounds of their 'clumsiness' and are even taking the place of men in fishing on Lake Titicaca. This example suggests that where women are able to take more control of their own lives they can improve their opportunities and social position.

As Rogers (1980) has shown, development agencies have tended to ignore women's productive roles and to superimpose patriarchal, Western views on non-Western societies. However, the impact of feminist ideas on policy in developing countries, combined with the declining economic position of many Third World countries, has helped to focus the thoughts of both national governments and international aid agencies on the role of women in development. At the same time, Third World women themselves are speaking out and demanding that women should be allowed to make decisions and to shape the societies in which they live. In practical terms this means a rethinking of the processes and priorities involved in development. For example, many women in developing countries would give a higher priority in development plans to having pure water piped to their houses than to building roads.

As geographers, we must be willing not only to look at new research paradigms and to query official data bases but also to study the implications of changes, such as those occurring on the shores of Lake Titicaca, for the distribution of spatial patterns of development. The flawed picture of female productivity provided by most national statistics has given development policy makers an implicit model of women's roles, but we should not assume that this model cannot change. We must examine more closely the regional variations first revealed by Boserup and the impact of time and distance constraints on the gender division of labour. Changes in traditional woman–environment relations are becoming widespread in the Third World and geographers have only just begun to recognize them.

Further reading

1 Essential reading for anyone concerned with women and development is E. Boserup's seminal text, *Women's role in Economic Development* (St Martin's Press, New York, 1970). A more recent examination of the role of women in the Third World and of the attitudes of development agencies to women is provided by B. Rogers, *The Domestication of Women* (Kogan Page, London, 1980).

2 For work by geographers the collection of papers edited by O. Horst, *Papers in Latin American Geography in honour of Lucia C. Harrison*, Conference of Latin Americanist Geographers, Special Publication no. 1 (Muncie, Indiana, 1981) provides a good introduction. This is the first book by geographers on Third World women.

3 For a view from those involved in actively changing their world we suggest M. Randall, *Sandino's Daughters* (Zed Press, London, 1981). This gives the testimonies of women involved in the Nicaraguan fighting.

Topics for discussion

1 A better understanding of the role of women in farming is essential if Third World food production is to be increased. Discuss.

2 Are women victims or beneficiaries of development?

PART III

Doing feminist geography

7 Feminism and methods of teaching and research in geography

In this chapter we look at what feminism has to say about how geography is taught and how geographical research is carried out. In the first section we examine women's low level of participation in professional academic geography. In the second section we will look more generally at the whole area of personal interaction between staff and students and within student groups to show how certain assumptions about female students may hinder their performance and restrict their opportunities. In the final section we will consider how a feminist approach to geography might alter the way in which we do our research.

We start with the issue of women's participation in studying and teaching geography because it raises questions (and provides some answers!) about why geographers have neglected or dismissed particular geographic topics, methods of research and ways of teaching which seem feminine or feminist. Certainly it is true that the majority of work both on 'the geography of women' and on 'feminist geography' (see Chapter 1, pp. 20–1) has been done by women rather than men. Therefore part of the explanation for the neglect of women's lives and of feminist issues seems to lie in the small number of women geographers. The poor representation of women in academic life also reflects the sexist nature of the wider social system within which we live. As feminists, who wish to change the patriarchal basis of our society, we feel we should also be trying to change what goes on in our own discipline.

Women and men in geography departments

Men are numerically dominant in almost all institutions teaching degree level geography, and also hold most of the positions of power. A survey was carried out by the IBG Women and Geography Study Group (WGSG) in 1980 of university, higher education and polytechnic departments offering geography degrees. Thirty-one out of forty-five university departments, sixteen out of twenty-seven polytechnic departments and three out of nine higher education college departments offering full-time degrees replied. Departments at the Open University and Birkbeck College, London University which offer primarily part-time degrees were also included in the survey. In only twelve of the forty-eight departments offering full-time degrees which replied did women undergraduates outnumber men (Figure 7). Although the part-time degrees offered by the Open University and Birkbeck should make these institutions more attractive to women with family responsibilities than conventional departments, they do not have an above average proportion of women students. However the survey did not ask questions about students' ages or family status.

The two departments in which women were most strongly represented were both in higher education colleges. Although there was considerable variability in the percentage of women the majority of departments had 40 to 49 per cent and overall 45 per cent of undergraduates were women. However, far fewer women than men go on to do postgraduate work: of the 315 full-time postgraduates listed by the surveyed departments only eighty-seven (28 per cent) were women. Perhaps one factor accounting for this low number of postgraduate women is the lack of women lecturing staff – women students, either consciously or unconsciously, may form the view that it is not a 'woman's career.' Women accounted for only 10 per cent of all the full-time

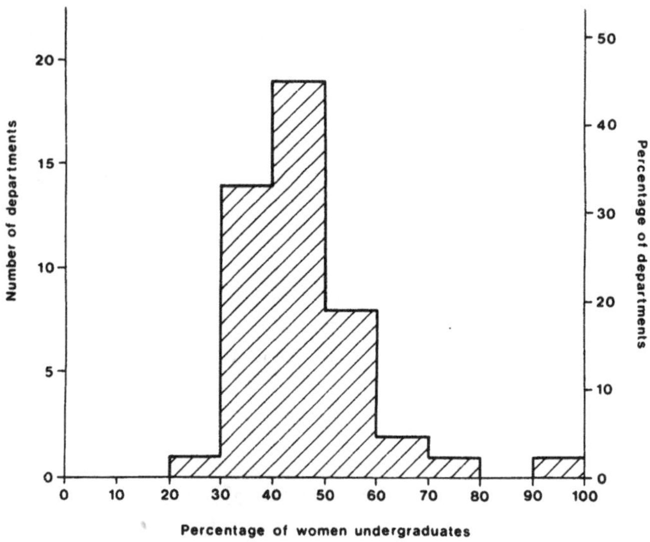

Figure 8 Women full-time undergraduates as a percentage of all full-time undergraduates in British geography departments offering higher degrees, 1980
Source: WGSG Survey (For details see text)

Table 1 *Number of women full-time lecturing staff in British polytechnic and university geography departments, 1980*

No. of women	*No. of departments*	*% of departments*	*cumulative %*
0	17	36	36
1	10	22	58
2	11	24	82
3	5	11	93
3+	3	7	100

lecturing staff in the departments surveyed. The survey found no full-time female lecturers in 36 per cent of the departments and only one or two in 46 per cent of departments, as is shown in Table 1. In no case did the proportion of women exceed one-third of all full-time staff.

Women were even more under-represented in senior positions at universities and polytechnics (senior lecturer and above at universities, principal lecturer and head of department at polytechnics) holding just 7 per cent of these senior positions; only four university professors were women and no polytechnic head of department was a woman.

Looking at the period 1933–82 Johnston and Brack (1983) have shown that the chances of a man becoming a reader (the next step down from professor in the academic hierarchy) were one in seven, but for a woman they were one in fourteen. Other positions of power are largely held by men. While in America the Annals of the Association of American Geographers (AAG) is now edited for the first time by a woman, the British equivalent, the Transactions of the Institute of British Geographers, has always been edited by a man. There has never been a female president of the IBG while only one of the present council members is a woman.

The situation in the United States and Canada resembles that in Britain, although the proportion of women undergraduates and faculty members in Canada is somewhat lower. In Canada in 1978–9, 38 per cent of undergraduates majoring in geography were women, while 28 per cent of postgraduates, 6 per cent of faculty members and only 3 per cent of tenured faculty were women. In the United States only 10 per cent of AAG members employed in colleges or universities were women and in 1979–80, 61 per cent of departments offering advanced degrees had no women faculty members. Only 22 per cent of the masters and doctoral theses completed in 1979 in the United States were by women (Zelinsky *et al.*, 1982; Henshall Momsen, 1980).

There is little information about the relative representation of women in the 'human' and 'physical' parts of the subject. More women are known to do arts subjects rather than sciences at both school and university, and we might therefore expect a higher proportion of women than men to prefer the 'soft' side of the discipline – social and human geography. However, figures from Reading University geography department – one of the three university departments that offer separate 'human', 'human and physical' and 'physical' geography degrees – belie this expectation. Students at Reading are free to choose their final degree after two terms of study. Figures for the years 1972–81 show that while more students of both sexes chose human geography in that period rather than pure physical geography, the percentage of women was *higher* among those doing physical geography, at 46 per cent, than it was among those doing human geography, for whom the figure was 42 per cent. In only two years out of the ten examined was the percentage of women students higher on the human than on the physical geography degree.

As far as postgraduate research on the human and physical sides of the discipline is concerned, few figures are available. One indication is given by looking at the sex of the authors of Ph.D theses as listed in the 'Geography Theses in preparation 1981-82, and completed 1981' put out by the Heads of Geography Departments Conference. For the 1981 academic year, twenty-three out of thirty-nine university and polytechnic departments in which Ph.Ds were completed identified the gender of successful candidates in their entry. More 'human' than 'physical' geography theses were written but the proportion of women was higher (25 per cent) among the physical geography researchers than the human geography reseachers, for whom the proportion was only 20 per cent. Unfortunately there are no comprehensive figures giving similar proportions among lecturing staff. Indeed, the whole question of the recruitment of women to study and research in geography and in its various subdisciplines needs further examination. However, the two cases quoted above do not suggest that women are biased against physical geography, although they re-emphasize the considerable fall in women's participation between undergraduate and graduate study.

There is little sign of any recent improvement in women's representation among postgraduate and lecturing staff in Britain despite the encouraging increase in female undergraduates. For universities we can compare the 1980 figures with those for 1978 collected by McDowell (1979). As Table 2 shows, although the percentage of women undergraduates has risen by 3 per cent (paralleling a general rise in the proportion of female undergraduates), the proportion of doctoral students has fallen by the same amount, while the proportion of lecturing staff shows little change. Figures for Canada and the United States over the last decade do show some gains in the number of graduate students and a very slight improvement in the status of women

Table 2 *Change in full time women students and lecturers in British universities, 1978–80*

	Year	
	1978	*1980*
First degree students – % female	41.9	45
Ph.D students – % female	31.0	28
Lecturers – % female	7.3	7.8

Source: McDowell (1979); unpublished WGSG survey (1980).

Table 3 *Percentage of university geography lecturers obtaining their first post who were female, 1933–80*

		Decade	
	1950s	*1960s*	*1970s*
Percentage female	11.6	8.5	7.9

Source: Johnston and Brack (1983).

faculty members. More disturbing are the figures produced by Johnston and Brack, which show that the proportion of women taking up university lecturing posts in Britain has been declining in the post-war period (Table 3). The figures suggest that women were not able to take advantage of the rapid expansion of universities which followed the Robbins Report, to the same extent as men.

Interpretations of women's low and possibly declining representation amongst geography postgraduate and lecturing staff will differ. Some may see it as confirmation of the view that fewer women than men are capable of the intellectual prowess such a position requires! We, however, consider it as evidence of the difficulties women face in entering a profession organized around a male way of life in a society in which women are socially and economically weaker than men.

We have already suggested that one factor deterring women from entering on an academic career is the lack of female 'role models'. Another factor which may disadvantage women trying to enter or gain promotion within the academic profession is their lack of entrée to the 'old boy' network. This may limit their opportunities in finding co-researchers and patrons (*sic*!) who can help by providing information on current research trends, funding and job opportunities (Figure 8).

With the growth of interest in feminism, and organizations such as the IBG Women and Geography Study Group in Britain, the AAG Committee on the Status of Women in Geography in the USA and the CAG Women and Geography Speciality Group in Canada, 'old girl' networks are beginning to be set up, but since so few women yet hold positions of power in the profession their influence is still weak. They can, however, help to remove the isolation felt by women who all work in male dominated departments. Eventually, we hope that women will be at least as well represented as teachers in higher education as they now are as undergraduates. Although an increase in the number of women will not in itself necessarily produce changes in the nature of academic life, feminists, both female and male, will try to alter the organization, structure and orientation of higher education and research to

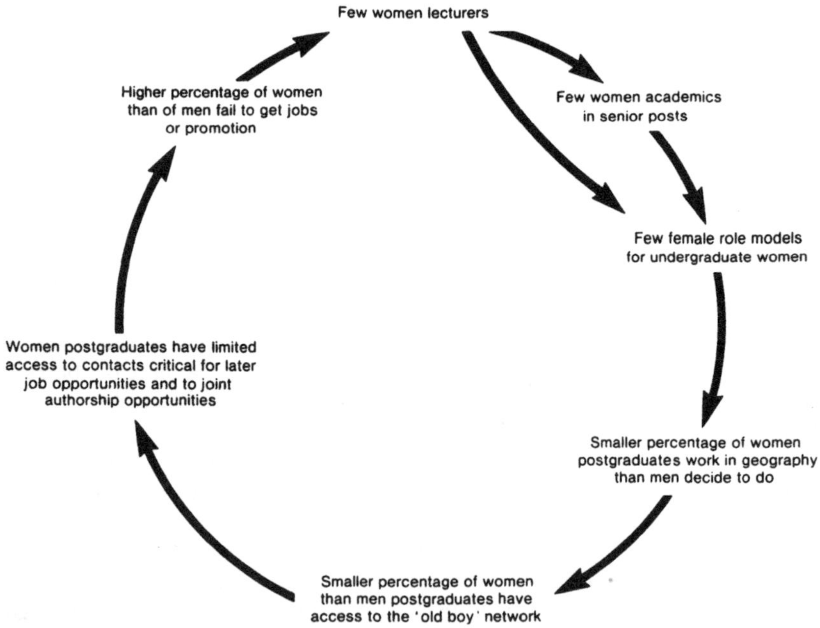

Figure 9 Possible chain of events explaining the low percentage of female academic geographers

Source: adapted from Christensen, 1978, and Zelinsky *et al.*, 1983

reflect general feminist goals. We consider that a more equal representation of women should help in achieving such far-reaching changes.

At present geography remains a male-centred subject, both in terms of the gender of those who structure its content through their teaching and research, and in terms of the nature of that content. In earlier chapters we have tried to show how feminist geographers have challenged the conventional view of what the subject is about; however, we have not considered the way in which a feminist approach may change not only *what* we study and teach, but also *how*. It is to these issues that we now turn.

We now want to examine the issue of personal interaction between geographers in schools, colleges, polytechnics and universities.

The processes of teaching and learning

We have suggested in the first section that academic geography presents a particularly 'male' image in the staff composition of individual departments and in the institutions of the discipline. In this section we examine the implications of this for the way in which geography is taught and for the manner in which female undergraduates are treated.

One of the respondents to the WGSG 1980 survey wrote:

> We treat men and women equally (I hope). . . . I personally find our women students (in general) more careful, conscientious and tidy than the men but perhaps a little less imaginative.

These are very common views concerning both the equality of treatment of male and female students and the characteristics of women students. They raise a number of important questions. First, *are* men and women treated equally? Should they be treated equally? Second, is this a true picture of women students? If so, why is this the case? If not, why is it such a commonly held view?

Little comparative research has been done into the teaching and degree performance of men and women undergraduates, and in geography there is almost none, so we have hardly any evidence with which to begin to answer these questions. However, a great deal of research *has* been done on school level teaching and on girls' upbringing. We can use this to suggest some interpretations which you may like to think about in relation to your own experiences, or perhaps to examine more rigorously through project work.

The body of research to which we have referred stresses the importance of the so-called 'hidden curriculum' – that is those things which are taught either unintentionally or without being made explicit, through the manner in which teaching is organized and performed. Different attitudes among teachers towards boys and girls, different assumptions about their future careers and different amounts of attention in and out of class have all been shown to be influential in moulding boys' and girls' expectations of themselves and in suggesting to girls, but not to boys, that a career is of secondary importance to marriage and bringing up a family.

How will such early training affect women's performance at degree level? Clearly it will discourage many from attempting to go on to higher education but for those who do go on to study for a degree, it does seem likely that many will indeed be 'careful, conscientious and tidy' and appear 'unimaginative'. However, as we have implied, these characteristics probably arise from parents, teachers and peers emphasizing, over many years, that quiet and conformist behaviour on the part of girls is 'good' while rowdy, non-conformist behaviour is 'bad'. As Table 4 indicates, boys in the United States are far more likely to be encouraged to be adventurous and non-conformist than girls. In the same context, Michelle Stanworth (1981), investigating the problems facing girls starting the 'A'-level syllabus in Great Britain, has stressed that unlike most boys, girls have to 'unlearn' the kinds of academic behaviour that had previously brought them success – that is conformity and correctness. Many never do 'unlearn' it at all before they begin their careers in higher education.

Another factor influencing women's academic achievement is that many women hide their abilities, even from themselves, since they find it difficult to cope with being 'brighter' than their male friends. They have learned early on that women are not supposed to outshine their men but should remain modestly half a pace behind! Such feelings encourage women to aim at jobs and academic qualifications below their potential and they will need positive encouragement to become more ambitious and self-assertive.

We have said that the 'hidden curriculum' is important at the school level; perhaps there are parts of the 'hidden curriculum' in higher education which also encourage women to underachieve. Lecturers' attitudes towards women students will influence the way in which they react towards shortcomings in their work. If lecturers believe that women students are generally competent but unimaginative, they may be more inclined to 'write-off' quiet women students, whereas quiet men students are more likely to be seen as having 'hidden potential'. While shyness is, of course, quite common among men, it is not reinforced by the belief that extrovert, socially aggressive behaviour is unmasculine, while such behaviour *is* commonly felt to be unfeminine. In mixed tutorial groups, therefore, women are more likely than men to be quiet and unforthcoming, particularly when they are in a minority. This lack of confidence over public speaking is common among women in many other situations and may well owe its origins to the 'hidden curriculum' of school, as well as to training at home. Tutors who are not aware of these difficulties may, after some initial attempts to help, simply assume (often incorrectly), that the student has nothing interesting to say.

Table 4 *Attributes of 'good' female and male pupils as viewed by American junior high-school teachers*

Adjectives describing good female pupils	*Adjectives describing good male pupils*
Appreciative	Active
Calm	Adventurous
Conscientious	Aggressive
Considerate	Assertive
Co-operative	Curious
Mannerly	Energetic
Poised	Enterprising
Sensitive	Frank
Dependable	Independent
Efficient	Inventive
Mature	
Obliging	
Thorough	

Source: Kemener (1965); quoted in Oakley (1981), p. 132.

A further, but related, problem is that both in everyday exchanges and in the classroom men or boys tend to dominate the conversation: they speak for longer and interrupt more than women (Spender, 1982). This seems to be true even where it is thought by those involved that everyone has a fair share of the talk. Similarly, researchers have found that schoolteachers give more time and attention (both positive and negative) to boys than to girls. Again this seems to occur without the teachers being aware of it and even when they are attempting to avoid doing it. Thus in higher education women are likely both to get and to demand less attention from teachers than men and to be more likely to feel insignificant and uninteresting. Once again such effects are likely to disadvantage girls in tutorials and seminars and to reduce their chances of being noted early on by lecturers as 'promising', 'two-one material' or 'a potential first'. An alternative danger faces those women who are intellectually adventurous and assertive and it is important to note that there *are* indeed such women students, despite the influence of early conditioning. Women who break the stereotype are often disliked by their peers and their teachers as 'aggressive', 'over-confident' or 'stroppy'.

Many people will feel that, while there may be some truth in the claims that staff do not notice clever girls early on or that they underestimate their ability, this is not very important because the prime method of judgement in higher education is assessment of written work, particularly written exams, where natural talent will be clearly evident to any marker. However, there is a difficulty with this argument. There is evidence to suggest that preconceptions about particular types of student can influence the marks given on written work. For example, Bradley (reported in Tysoe, 1982) compared the marks given to degree level project work by assessors who knew the students well because they had supervised the project, assessors who did not know the students but knew their sex, and assessors who knew neither. She found that, compared with the entirely independent assessors, there was a significant tendency for the second group to give the women middle range marks while the men were marked both more severely and more generously. In other words, the markers' preconceptions about the kinds of results women would get influenced their judgement of the work itself. Among the project supervisors who knew the students very well this stereotyping did not occur. Since in most written exam marking the lecturer does not know the student very well, preconceptions about particular types of student may influence the marking. Although the most and least academically able women would probably not be affected by such bias, the majority of students do not fall into these categories and systematic marking towards the middle could result in more women getting lower seconds and fewer getting firsts than their performance warrants. Clearly, there is a strong case for exams to be marked without the student's name or sex being known – a practice which is followed at a few, but not at most, universities.

An important point that emerges from the discussion so far is that lecturers are frequently unaware that their ways of behaving and prejudices may disadvantage women. They may well believe, and think that they ensure, that their judgements of intellectual ability are unbiased and that once on a degree course women have equal opportunities to their male peers. There are also some lecturers whose behaviour is more overtly sexist and objectionable – teasing of women students, suggestive remarks and outright physical harassment are not unknown. Such behaviour among male undergraduates also occurs and can certainly affect women's participation in group work either in seminars or on field-work. Where there are quite a few women in a class, particularly if they can get together, such behaviour is perhaps easier to combat (because easier to identify) than the more subtle forms of bias. Here, we must recognize that all of us are open to the influence of stereotyped views and all of us, men and women, staff and students, need to be more sensitive to the effects of our own behaviour.

So far we have concentrated on the problems of interaction with lecturers and other students that women may face. However, the organization of academic life may also pose practical difficulties for women, and for parents of young and school-age children. Three examples may help to explain this point. First, in Britain, the fixed full-time three year degree course with lectures between 9 a.m. and 6 p.m. poses particular difficulties for students with children, since school hours run from 9 a.m. to 3.30 p.m. or 4 p.m. and school half-term holidays may last a week and take place during university terms. Secondly, the importance of compulsory one- or two-weekly field courses in geography creates problems for both students and staff members with young children. Finally, women students may run the risk of physical attack. On many campuses women may be deterred from using the library in the winter evenings for fear of attack when walking home at night. There are also dangers involved in undertaking field-work alone, whether in isolated rural areas or in towns and cities. While this is not a problem specific to geography students, it is an issue on which their skills can be turned to good account, as at least two undergraduate projects on the topic have shown (Hibberd and Hindle, 1981; Townsend, 1982).

The last example above brings us to the question of what can be done about the problems we have outlined. Perhaps the most important step is for individual women to realize that they are not alone in their doubts and difficulties and that their problems are not simply a result of their own inadequacies. Discussion with other women can help to sort out ways of dealing with such problems as nervousness in speaking or sexist behaviour from staff or students. Getting together with others, both women and men, is also important if we want to change the organization and content of teaching – the nature of the curriculum, the timing of classes, the structure of degree courses. We need to work out alternative ways of organizing our teaching and together press for changes to be implemented.

Doing feminist geography

In this section we focus on *project and thesis work* since projects, undergraduate theses, or regional or extended essays, are usually a fairly substantial component of a geography degree. We want to show how an appreciation of the issues raised by feminism may alter how geographers set about academic research.

Awareness of feminist ideas should alter the topics selected, the theoretical context within which empirical data are collected and interpreted, and the methods of data collection and analysis employed. The most obvious way in which projects might reflect a feminist approach to geography is for students to undertake work which critically examines the ways in which women's opportunities are structured by the spatial distribution of goods and resources.

We hope that some of the topics we have included at the end of this chapter will provide useful starting points for such work. But there are many other issues to be explored. For example, in some of the earlier chapters we have used existing evidence to show how the built environment, the structure of the housing market, the classification of job skills and employment location, the ideology of development agencies and other social institutions are based on and create a patriarchal view of women's place in the world. We need more empirical research in different parts of the world and in different localities to explore and expand our knowledge of the working-out of patriarchal assumptions in space. The specific nature of the relationships between domestic work and waged work, and the rolle of women within the family and within the local community all vary across time and space and require analysis through locality-based studies.

Compare, for example, the different conceptions of women's roles in areas of the United Kingdom such as Lancashire and the North-East; the first with a long tradition of female employment in the textile industry and participation in work-based activities such as trade unions, the latter with, until recently, a predominantly male-employing industrial structure and few paid jobs for women. Coal-mining and ship building involve heavy, dirty work, often on a shift basis, and women traditionally were expected to be at home to provide meals and wash clothes. Their predominantly domestic role is reflected in the local culture based around working men's clubs and other exclusively male activities. In this area, in contrast with Lancashire, few women are involved in work-based activities or local politics.

There are many other topics on which interesting project work relating to women might be carried out: access to child care, to hospitals and clinics, women's role in production and exchange in developing countries and so on. But we hope that in the future feminist geography will not focus exclusively *on women*, nor be undertaken solely *by women*. Feminist geographers are concerned with the structure of social and spatial relations that contributes to women's oppression. Consequently men too must be included in the analysis: as oppressors in certain situations, as oppressed in others. Moreover, men as well as women can be feminists, concerned about gender roles and inequality, who reflect this concern in their research. While in this book we have tended to focus on the neglect of women and of 'women's issues' in traditional geographic research, geographers have also ignored the gender basis of *male* behaviour. A truly feminist geography thus embraces the gender roles of women and men (Tolson, 1977). Eventually, it will also need to transcend the common disciplinary divisions such as those between 'industrial geography' and 'urban geography'. However, as this book is an introductory text, we have deliberately chosen to remain within these compartments in order to present a more readily comprehensible critique of geography as it is currently taught in schools and higher educational institutions.

We suggested above that doing feminist research involves changes not only in research content but also in research method. We need to reassess conventional ideas about the conduct of research. Standard texts on research methods which rely on positivist techniques stress the importance of 'objectivity', by which they mean a particular form of disinterestedness considered necessary for 'unbiased' access to 'the truth'. Thus they warn of the dangers of personal involvement with the subjects in questionnaire and interview research. However such non-involvement can itself obscure truth. We need to allow the purportedly 'female' characteristics of empathy, intuition and concern for the feelings and emotions of others to temper the purportedly 'male' characteristics of objectivity and rationality if we are to understand people's actions and attitudes. The objectivity (in the sense of 'truth') of such 'soft' approaches has already been vigorously defended in other social science disciplines (for example, see R. Harre, 1979). These types of issues loom largest on the 'human' side of the discipline where interview and questionnaire techniques are more commonly employed. They are beginning to be recognized not only by feminists but by those working in the 'humanistic' tradition, who use 'soft' data and qualitative as well as quantitative techniques.

Gender relations are important in questionnaire and interview design. For example, the attitude and gender of an interviewer inevitably affect the response of the interviewee. Male and female interviewers will not necessarily see or be allowed to see the same social worlds. A male student, for example, would probably be ill-advised to undertake work on the problems of access to antenatal care of women from minority groups, or at the least he should expect to be told very different things from those told to female students. Age, nationality, social class and family status too will affect the results. These issues are already well known to those designing questionnaires who frequently recommend that interviewers should be 'pleasant' middle-aged women because they are not thought to be perceived by respondents as 'threatening'. Researchers need not only to recognize such aspects of gender relations in research design but also to make a critical examination of their impact an integral part of research projects.

Taking gender relations into account raises other problems for project work. Often the data are simply not available or are classified and presented in inappropriate ways. The official manual of interview techniques used by the Office of Population Censuses and Surveys defines men as the heads in households containing men and women. Until 1981, the official Census form did not even allow for the woman in a conventional married couple to be the joint head of the household. Often it is only the man who is interviewed, perhaps from quite reasonable desires to reduce the time and cost of the survey, but consequently power relations within the households are ignored and certain types of information – on spending habits for example – may not be accurate. For example, many farmer's wives play a major economic role on the farm but this is ignored in many farm studies which consider them only in terms of their household role (Gasson, 1982). In dual-headed households, the data collected may vary depending on whether or not both partners were interviewed and whether they were questioned separately or together.

Collecting the data is not the sole problem. Conventional methods of assigning a social class category to a respondent are based on male employment. Women are sometimes ignored or, if married or living at home, they are usually classified according to their husband's or father's occupation. Thus it is often impossible to compare the social class of single with married women, and women living alone with those still in their parental home. In historical geography, the problems are even greater as methods of inquiry have to be devised which rediscover the lives of those women who were 'hidden from history': wives and servants of influential men, for example. In certain cases the information is just not available at all. These types of problems, of course, have much in common with those raised by research on the working class as a whole. Historical information about the lives of oppressed groups is often difficult to find. These are issues of which we should be aware when designing our own research proposals or attempting to re-use statistics collected by others.

Official statistics produced by the government are often particularly difficult to use as they are based in many cases on 'common sense' conceptions of what is appropriate. The *Fertility Tables*, for example, seldom include information about men. But why not? Data about fathers surely are relevant. The *New Earnings Survey* includes details of part-time employment among women but not among men. It is not only women who work part-time, although it is more usual. In the *Family Intentions Survey* only women were interviewed about intended family size. Other problems arise, particularly for geographers, as many official statistics, when broken down by standard region, omit the information for women. It is important that feminist geographers publicize these problems and join the working groups set up within our discipline to advise and assist in the production of new classifications and tabulations of official statistics. As in the area of teaching, we need to work together to improve and change both the content and the methods of our work.

Further reading

1 A very thought-provoking book on research methods is H. Roberts, (ed.), *Doing Feminist Research* (Routledge and Kegan Paul, London, 1981). Although the contributors to this book are sociologists, a great deal of what they have to say about research methods, interviewing, personal interaction and data collection is equally applicable to geography.
A. Oakley and R. Oakley, 'Sexism in official statistics' in J. Irvine, I. Miles

and J. Evans (eds.) *Demystifying Social Statistics* (Pluto Press, London, 1979) provides a good discussion of the limitations of government statistics. Other chapters in this book may also be of interest to those who feel that statistics and statistical techniques hide as much as they reveal.

2 On teaching and learning, M. Stanworth, *Gender and Schooling* (Hutchinson and the Explorations in Feminism Collective, London, 1983) provides a very useful short treatment of school teaching. Another stimulating book is D. Spender, *Invisible Women: the schooling scandal* (Writers' and Readers' Publishing Co-operative, London, 1982).

Project topics

We have listed below a number of project topics linked to the material covered in the four chapters in Part II of the book. They are intended to be suitable either for group work in a class project or for undergraduate thesis research. We have not given detailed outlines of how the suggested topics might be carried out – our intention is to give you ideas about the kinds of research work that can be done, with some hints on methods of approach. By linking the projects to the material in Part II, we intend to show how topics for project work can be related to and arise from a feminist perspective on geographical issues. There is a need for feminist research in all areas of human geography and we hope some of you will apply the ideas in this book to projects in fields other than those discussed here.

Urban Spatial Structure

1 Study the development of a small suburban area in your locality in *either* the late nineteenth century *or* the twentieth century in relation to the issues discussed in Chapter 3. Select a particular aspect of its development for study: for example, for a middle-class suburb in the nineteenth century examine the pattern of domestic service employment; for a working-class suburb look at changes in the nature and, if possible, location, of female and male employment; find out how the provision of local facilities such as schools, shops, health care, employment and transport has changed. Use contemporary local newspapers to gain an impression of women's and men's lives in the area and the attitude of local businessmen, politicians and people of influence to its development.

If you are studying a nineteenth-century suburb you can look at original census returns for 1861, 1871 and 1881. Many local libraries have facsimile copies of returns for their area, otherwise you have to go to London, to St Catherine's House, OPCS, Kingsway, WC2, to look at the original returns.

Census data on more recent suburbs is only available at Enumeration District or Ward level, and you would need to check that the suburb you wish to study is identifiable in the available census data (see Short, 1980).

2 Examine the planning documentation on a fairly recently developed public housing estate in your area and interview the local authority planners and housing officers to discover their attitude to provision for women's needs. Compare this with your findings from an interview survey of the opinions and experiences of women living on the estate.

3 Interview those involved in women's groups (of all types) and other pressure groups in your area concerned with *either* child care *or* care for the elderly to see in what physical and social form they think this care should, ideally, be provided. Compare your findings with the ideas of different writers on non-sexist environments to which we referred in Chapter 3. The strategy for such a survey would be difficult to devise and you would need to have some preliminary unstructured discussions with a few members of such groups before you decided on your approach. Among the survey techniques you might consider are: unstructured individual interviews; tape-recorded group discussions with a loosely-structured format; using a mixture of structured and unstructured questions.

Women's employment, industrial location and regional change

1 Investigate the relationship between the gender and skill composition of the labour force in your local area and the current industrial structure. Examine how the composition and types of industry in your area have changed since the nineteenth century. Can you link these changes to the changing urban spatial structure?

2 Find out which are the major female employing manufacturing firms in your area. Ascertain why the firms are located in this particular area, why they prefer employing female workers, and what kind of work the women do. Approach a few local plant managers with a short questionnaire and try to arrange a visit round the shop floor. Comment on what you are shown and/or discuss this with a union representative. Relate your findings to the ideas discussed in Chapter 4.

3 As we have emphasized, most women work in the service sector. Select a major public or private service sector employer in your region – i.e. one providing distribution, insurance, banking, cleaning, catering or transport services, or involved in public administration. Find out the geographical area(s) served, what services are provided and the number of women and men employed. Compare the jobs done by men and women and assess the extent to which the employer's demand for types of workers such as male or female, part-time, clerical, skilled, unskilled is concentrated in particular towns or locations within them. Consider the implications for women's employment opportunities.

4 Find a sample of working women (go to a few workplaces, such as local shops, factories or hospitals, and ask for volunteers). Find out what jobs they do, where their work is relative to their home. Find out how they get to and from work. Plot a week's activities for these women and discuss the links between the sorts of paid jobs done by women and their domestic role at home. What changes do you think need to be made (if any) in the organization of domestic and waged work?

Access to facilities

1 Investigate the level of awareness of women concerning the services available in their local environment. Consider how information might be better disseminated. (This could be carried out mainly through a questionnaire survey.)

2 For your District Health Authority find out what types of maternity care are provided. Examine the problems that physical accessibility poses for women and how it affects their chance of maternity care. (Interviews with women recipients of maternity care would have to be supplemented by information collected on the provision of care.)

3 Conduct a survey to investigate women's attitudes towards, and use of, the health service. Consider, in particular, the constraints which women face in using health facilities and their opinions on the adequacy of the services provided. (This could be carried out as a group project with each student interviewing a set number of respondents after the survey has been designed jointly.)

Women and development

Because of the difficulties for most students in carrying out project work in Third World countries we have not suggested project topics on the same lines as those linked to earlier chapters. Instead we have provided some data analysis and interpretation exercises. We hope these will give you ideas both about the kind of re-interpretation of secondary data that can be done and about the kind of field data that can be collected.

The constraints of time on Third World women

Although the details of women's roles may differ from culture to culture, women carry out a multiplicity of tasks in traditional societies, which may be broadly grouped into breeding and feeding roles. Because of this Boulding (1977, p. 111) suggests that:

> the special set of spaces in every society where women carry out their productive roles, be referred to as the *fifth world*, which is found on every continent: in the family farms and kitchen gardens, the nursery and the kitchen. Within the rural and non-industrial parts of the fifth world, women breed babies, produce milk to feed them, grow food and process it, provide water, fuel and clothing, build houses, make and repair roads, serve as beasts of burden, and sit in the markets to sell the surpluses.

The amount of time women spend on any one type of task will vary with the seasons, the age of family members and the life cycle of the woman.

Mueller (1982) suggested that in developing countries a woman's leisure time is the first element to be sacrificed whenever market work, agriculture or child care demand a lot of time. Spiro (1981) showed in her work in Nigeria that the amount of time women spent each week in marketing, food processing and farming varied with the annual crop cycle. Yet Harry's (1980) work in Trinidad revealed no significant differences in the hours per day worked by men and women farmers throughout the year. On average men worked 4.9 days per week while women worked 4.8 days, although there were sectoral differences with women working longer hours than men on the rice and dairy farms. In Nevis, on the other hand, women spent significantly fewer hours than men, working only 4.8 days per week on the farm while men put in 5.5 days. Seasonal differences in the sexual division of labour were also more apparent among women farmers in Nevis, with the weekly hours worked by females being equivalent to 72 per cent of male hours in the busy season and only 66 per cent in the quiet season.

The workload of Caribbean women smallholders is very heavy. They often work a 15-hour day beginning by preparing the family breakfast as early as three or four o'clock in the morning. Both men and women work on the farm in the morning and afternoon with a rest period in the middle of the day. In Nevis women are more likely to spend the afternoon on non-agricultural work than in Trinidad, and Nevisian men work both earlier and later than Trinidadians

Table 5 *Sexul division of labour on the farm by time of day worked*

Time of day		*Percentage of workers*			
		Nevis		Trinidad	
		Male	*Female*	*Male*	*Female*
Very early	(4.00 to 6.00)	12	7	0	0
Early morning	(6.00 to 8.00)	63	48	55	77
Mid-morning	(8.00 to 10.00)	68	68	72	72
Late morning	(10.00 to 12.00)	44	30	60	61
Mid-day	(12.00 to 14.00)	16	20	30	57
Mid-afternoon	(14.00 to 16.00)	37	21	52	47
Late afternoon	(16.00 to 18.00)	35	25	22	18
Early evening	(18.00 to 20.00)	1	0	0	0
		N = 71	N = 28	N = 116	N = 114

Sources: For Nevis – fieldwork, 1979; for Trinidad – adapted from Harry (1980), Table 12.

Table 6 *Agricultural problems*

	Montserrat				*St Lucia*			
	Male		Female		Male		Female	
Problem	*% farmers mentioning problem*	*Rank of problem*	*% farmers mentioning problem*	*Rank of problem*	*% farmers mentioning problem*	*Rank of problem*	*% farmers mentioning problem*	*Rank of problem*
Difficulties in obtaining labour	27	5	64	5	58	2	80	1
Road & transport	23	3	18	7	36	1	60	2
Marketing	9	6	9	8	58	5	50	4
Plant pests & diseases	55	7	73	6	68	6	60	6
Water shortage	41	4	36	3	38	4	40	3
Soil erosion	16	10	14	10	38	10	50	5
Difficulties obtaining loans	16	3	0	–	22	3	0	–
Shortage of land	23	9	18	5	12	7	0	–
Land tenure	5	12	0	–	6	9	10	7
Old age & sickness	5	2	14	1	6	11	0	–
Rocks & stones in soil	11	7	5	5	0	–	0	–
Floods	0	–	0	–	6	8	0	–
Drought	27	1	27	4	0	–	0	–
Livestock damage to crops	5	3	14	2	0	–	0	–
Mechanization	2.5	8	5	9	0	–	0	–
Theft	2.5	8	0	–	0	–	0	–
No problems	9		0		2		0	–

Source: Field surveys.

(Table 5). These contrasts reflect both cultural differences between the East Indian farmers of Trinidad and the Afro-Caribbean farmers of Nevis, and the greater importance of off-farm work to Nevisians.

About one-third of West Indian smallholdings are operated by women. Women farm operators are found most commonly in households headed by women, although women may also run the farm when their male partner has a full-time job off the farm. Farms operated by women tend to be smaller, more isolated and to have poorer soils than those of men farmers. Land farmed by women has usually been inherited and is held freehold or as family land. It is used predominantly for subsistence production of food crops rather than for commercial farming. These characteristics suggest that farming is seen by women as a way of making use of the resources available rather than as a chosen way of life. A comparison of the problems faced by male and female farmers in Montserrat, where women made up 55 per cent of the small farmers, highlights some of the differences between the sexes in their agricultural roles (Table 6). The overall picture of female-operated farms in the West Indies is one of marginality in terms of capital, land and labour resources.

Exercises

1 Using Table 5, compare and contrast the work patterns of men and women in Trinidad and Nevis. Suggest reasons for these differences. Indicate the expected impact on women's farm work hours of:
   (a) a full-time off-farm job for the male partner
   (b) the presence of school-age children in the family
   (c) the presence of an adult daughter in the household

2 Using Table 6, suggest reasons for the different ranking of farming problems between the sexes. What light does this table throw on the characteristics of male- and female-operated farms?

References

Boulding, E., *Women in the Twentieth Century World* (John Wiley, New York, 1977).

Harry, I. S., 'Women in agriculture in Trinidad', Unpublished M.Sc thesis, (University of Calgary, Canada, 1980).

Henshall Momsen, J. and Townsend, J. (eds.), *Women's role in Changing the face of the developing world,* IBG Women and Geography Study Group (Geography Dept, Durham University, 1984).

Henshall, J. D., 'Women and small scale farming in the Caribbean', in O. Horst (ed.), *Papers in Latin American Geography in Honor of Lucia C. Harrison* (CLAG Publications, Muncie, Indiana, 1981), pp. 44–56.

Mueller, E., 'The allocation of women's time and its relation to fertility', in R. Anker, M. Buvinic and N. H. Youssef (eds.), *Women's Roles and Population Trends in the Third World* (Croom Helm, London, 1982), pp. 55–86.
Spiro, H. M. 'The Fifth World: Women's rural activities and time budgets in Nigeria', Occasional Paper no. 19 (Dept of Geography, Queen Mary College, University of London, 1981).

8 Conclusion

In the first chapter of this book, we argued that one reason why geographers should 'bother' with feminism was that a recognition of the significance of gender relations in shaping the world in which we live would improve our understanding of that world. We hope that the four examples of feminist work we presented in Part II of the book have begun to show how focusing on gender relations changes and improves existing geographic analyses. However, we also argued that feminist geography must be centrally concerned with *changing gender relations* (and not merely with understanding them) in order to bring about equality between women and men. Perhaps some of you will be sceptical about the relevance of such 'academic' enquiries to more direct political movements which attempt to alter our society. While academic research certainly can be a way of diverting attention from the need for action, it is not inevitably so. We ask sceptics to consider the following arguments. Firstly, we certainly hope that few who have read the material in this book and in our references remain unconvinced that women do indeed suffer from a lack of power, and are unequal in both the developed and developing world. Thus, by changing the terms in which people understand and interpret the world, feminist studies can help to bring about a wider recognition of patriarchal relations and gender inequalities *and* of the need to change them. Secondly, the need for 'theoretical' or 'academic' feminist analyses arises out of the practical problems that feminists face in trying to change the current patterns of gender inequality. For example, the failure of legislation to bring about equality at work has prompted studies which examine, in depth, the barriers to improvement that the legislation has failed to overcome. Similarly, women's experience of the inadequacy of health care and the difficulties of changing it has been an important stimulus to feminist analyses of current health care. In so far as these studies help to show *where* change is needed and *how* it might be brought about, they become part of the armoury with which feminists fight for social change. Moreover, as we emphasized in Chapter 7, the world of higher education is itself a part of society which both reflects and reinforces existing social relations. As feminists we must therefore struggle to change the aims and practice of our learning, teaching and research as well as seeking change elsewhere. The experience of such struggles may well help us to better understand how to achieve wider social changes.

If feminist 'theory' is concerned with helping to bring about social change then it follows that we must consider the implications for social change of the research that we do, and this will clearly influence the kinds of research questions that we ask. It should also influence the research methods that we employ so that we do research not *on* people but *with* them. These points can be made about feminist work in any discipline, but as geographers we need to consider whether there is any specific contribution that we can make.

The research reported in Part II of this book gives some indication of what that contribution could be. Work on how the location and spatial organization of different types of employment, services and residential areas affect and are affected by women's and men's contrasting power and social roles, is clearly of relevance to attempts to improve women's position. In Part II we also discussed variations in 'patriarchal relations' through time and between non-capitalist and capitalist societies, but we said rather less about variations in gender relations between regions or local areas. This is still an under-researched area, although such variations certainly exist and must be investigated if we are to develop our understanding of the ways in which gender, race and class relations interact. It is here perhaps that geographers may make an important contribution to the understanding of patriarchal relations. It is clear that relations between men and women vary in different areas within one country. The past history and previous experiences of each region in, for example, the type and nature of alternative employment opportunities, in patterns of migration, in the nature of the housing market, affect the particular ways in which gender divisions in the workplace, in the wider community and within the home, operate at the regional and at the local level. Thus there is a need for more case-studies of changes in interactions and interrelations at the local level. Such case-studies should investigate women's and men's everyday experience of, and action towards, their environment from a perspective which examines the links between those experiences and actions and wider structures of social relations. We believe, therefore, that geographers have a contribution to make to feminist theory and practice by drawing attention to significant, but often neglected, variations in space.

Such local case studies also exemplify one of the main challenges that feminism poses to the practice of academic geography. This is the need to eventually break out of the established subdivisions or specialisms of contemporary geography. Although in this introductory book we decided to look in turn at cities, work, access to facilities and the development process, we have tried to show you where the links between these areas of research exist. For example, we have argued that the experience of women in the labour market cannot be understood without an analysis of gender divisions within the home and, indeed, vice versa. Thus, the tendency to specialize in teaching and in research on industrial geography, urban geography, or Third World studies means that certain sets of social and spatial relations are in danger of being overlooked.

Finally, we hope that you have enjoyed reading this short book and exploring some of the growing literature on feminism and geography. We hope that it will be a continuing support, both academically and practically, in the future. Because feminism challenges us to examine our personal, political and academic lives, it is bound to sometimes be a painful as well as a rewarding path to follow.

Bibliography

Aldred C. (1981) *Women at Work*. Pan Trade Union Studies, London.

Alexander S. (1980) 'Introduction' in Herzog M. *Frcm Hand to Mouth: Women and Piecework*. Penguin, Harmondsworth.

Allin P. (1982) 'Women's activity rates and regional employment markets' Paper presented to: *British Society for Population Studies, Conference on Population change and Regional Labour Markets* OPCS *Occasional Paper no.* 28, HMSO.

Alonso, W. (1964) *Location and Land Use*, Harvard University Press, Cambridge, Mass.

Ashford J. (1981) 'Trends in maternity care in England and Wales 1963 – 1977' in McLachan G. (ed), *Matters of Moment*. OUP, Oxford.

Austrin T. and Benyon H. (1979) 'Global Outposts: The working class Experience of Big Business in the North-East of England, 1964 – 1979'. *Discussion document*: Dept. of Sociology, University of Durham.

Bagchi D. (1981) 'Women in Agrarian transition in India: Impact of Development'. *Geografiska Annaler*, **63B**, 109-117.

Bale J. (1982) 'Sexism in geographic education' in *Bias in Geographical Education*. Department of Geography, University of London Institute of Education.

Barrett, M. and McIntosh, M. (1982) *The anti-social family*. Verso Editions, London.

Barrett M. and Roberts H. (1978) 'Doctors and their patients: The social control of women in general practice', in Smart C. and Smart B. (eds) *Women, Sexuality and Social Control*. Routledge and Kegan Paul, London.

Bassett K. and Short J. (1980) *Housing and Residential Structure: alternative approaches*. Routledge and Kegan Paul, London.

Beechey V. (1978) 'Women and production: a critical analysis of some sociological theories of women's work' in Kuhn A. and Wolpe A. (eds), *Materialism and Feminism: Women and Modes of Production*. Routledge and Kegan Paul, London.

Benton J. (1984) 'The changing position of Aymara women in Bolivia's Lake Titicaca region', in Townsend J. and Momsen J. (eds), *Woman's Role in Changing the Face of the Earth*. IBG Women and Geography Study Group.

Beveridge Report (1942) *Report on the Social Insurance and Allied Services*, Cmnd 6404, Her Majesty's Stationery Office, London.

Boserup E. (1970) *Women's Role in Economic Development*. St Martin's Press, New York.

Boston S. (1980) *Women Workers and the Trade Unions*. Davis-Poynter, London.

Boulding E. (1977) *Women in the Twentieth Century World*. John Wiley, New York.

Bourne, L.S. (1982) (ed), *Internal structure of the city: readings on urban form, growth and policy*. 2nd edition, Oxford University Press, New York.

Bournville Village Trust (1941), *When We Build Again*.

Bowers J. (1970) 'The anatomy of regional activity rates', in *National Institute of Economic and Social Research. Paper 1*. Cambridge University Press, Cambridge.
Bowlby S.R., Foord J., Lewis J. and McDowell L. (1983) 'Recession and the spatial and sexual division of labour: the case of Women's employment in New Towns in the Development Areas'. Paper presented to SSRC conference on Urban Austerity, Urban Change and Conflict, Clacton
Bowlby S.R, Foord J. and MacKenzie S. (1982) 'Feminism and geography', *Area*, **14**(1), 19-25.
Brietenbart E. (1982) *Women Workers in Scotland*. Pressgang, Glasgow.
Bruegal I. (1979) 'Women as a reserve army of labour'. *Feminist Review*, **3**.
Burgess E.W. (1925) 'The growth of the city' in Park, R.E., Burgess, E.W. and McKenzie R.D. (eds), *The City*, University of Chicago Press, Chicago.
Buttimer A. (1972) 'Social space and the planning of residential areas', *Environment and Behaviour*, **4**, 279-318.
Buttimer A. (1976a) 'Grasping the dynanism of lifeworld', *Annals Association Amer. Geography*, **66**, 277-292.
Buttimer A. (1976b) 'Beyond sexist rhetoric: horizons for human becoming' in Burnett P. (ed), *Women in Society: A New Perspective* (mimeo).
Buvinic M. and Youssef N.H. (1978) *Women-headed households: the ignored factor in development planning*. Center for Research on Women, Washington D.C.
Carlstein T., Parkes D. and Thrift N. (eds) (1979) Timing space and spacing Time *Socio-Economic System Vol. 2 Human Activity and Time Geography*. Edward Arnold, London.
Carney J., Hudson R. and Lewis J. (eds) (1980) *Regions in crisis: New Perspectives in European Regional Theory*. Croom Helm, London.
Carney J. (1980) 'Regions in crisis: accumulation, regional problems and crisis formation', in Carney J., Hudson R. and Lewis J. (eds) *Regions in crisis: New Perspectives in European Regional Theory*. Croom Helm, London.
Castells, M. (1977) *The Urban Question: A Marxist Approach*, Edward Arnold, London.
Cavendish R. (1982) *Women on the Line*. Routledge and Kegan Paul, London.
Christopherson S. (1980) 'Labour Intensive Industry and Third World Urbanization: The case of the Maquiladoras in Juarez'. Paper presented at the Annual Meeting of the Association of American Geographers, Louisville, Kentucky.
Clarke J. (1983) 'Sexism, feminism and medicalism: a decade review of literature on gender and illness', *Sociology of Health and Illness*, 5.
Coates B.E., Johnston R.J. and Knox P. (1977) *Geography and Inequality*. OUP, Oxford.
Cockburn C. (1983) *Brothers: Male Dominance and Technological Change*. Pluto Press, London.
Cooke P. (1981) 'Tertiarization and socio-spatial differentiation in Wales', *Geoforum*, **12** (4)
Coote A. and Campbell B. (1982) *The struggle for Women's Liberation*. Pan Books, London.
Coote A. and Campbell B. (1982) *Sweet Freedom*. Picador, London.
Coote A. and Kellner P. (1980) 'The working women her job her politics and her union', in *Hear This Brother: Women-Workers and Union Power*. New Statesman Report No.1, London.

Counter Information Service (1978) *Women under Attack: Anti-Report No.15*. CIS, London.
Counter Information Service (1981) *Women in the Eighties: Anti-Report No.28*. CIS, London.
Coupland V. (1982) 'Gender class and space as accessibility constraints for women with young children', in *Contemporary Perspectives on Health and Health Care*. Health Research Group, Queen Mary College, London.
Coyle A. (1982) 'Sex and skill in the organization of the clothing industry', in West J. (ed), *Work, Women and the Labour Market*. Routledge and Kegan Paul, London.
Christensen K.E. (1979) 'Geography and Space: an epistemological dilemma and resolution', Unpublished Masters thesis, Pennsylvania State University.
Croll E.J. (1979) 'Socialist Development Experience: Women in Rural Production and Reproduction in the Soviet Union, China, Cuba and Tanzania'. Institute of Development Studies, University of Sussex, Discussion Paper No. 143.
Damette F. (1980) 'The regional framework of monopoly exploitation', in Carney J., Hudson R. and Lewis J. (eds) *Regions in Crisis: New Perspectives in European Regional Theory*. Croom Helm, London.
Davidoff L., L'Esperance and Newby H. (1976) 'Landscape with figures: home and community in English society', in Mitchell, J. and Oakley A. (eds) *The Rights and Wrongs of women*. Penguin, Harmondsworth.
Davis A. (1982) *Women, Race and Class*. Women's Press, London.
DeBrettville S.L. (1981) 'The woman's building: physical forms and social implications', in Keller S. (ed) *Building for Women*. D.C. Heath and Co, Lexington, Mass.
Deere C.D. and Leon de Leal M. (1981) 'Peasant Production, Proletarianization and the Sexual Division of Labour in the Andes', *Signs*, 7.
Dicken P. and Lloyd P.E. (1981) *Modern Western Society: a geographical perspective on work, home and well-being*, Harper and Row, London
Donovan J. (1983) *Black people's health: a different way forward ?*, IBG Medical Study Group Conference, Leeds, 1983.
Dunford M. (1979) 'Capital accumulation and regional development in France' *Geoforum*, **10**.
Dunford M., Geddes M. and Perrons D. (1981) 'Regional policy and the crisis in the U.K.: a long term perspective', *International Journal of Urban and Regional Research*, 5.
Engels F. (1968) 'Origin of the family, private property and the state', in Marx K. and Engels F. *Selected Works*. Lawrence and Wishart, London.
Eyles J., Smith D. and Woods K. (1982) 'Spatial resource allocation and state practice: the case of health service planning in London', *Regional Studies*, **16**.
Eyles J. and Woods K. (1983) *The Social Geography of Medicine and Health*, Croom Helm, London.
Feminist Anthology Collective (1981) *No Turning Back: Writings from the WLM 1975–1980*. The Women's Press, London.
Firestone S. (1971) *The Dialectic of Sex*. Bantam, New York, Toronto & London.
Firn J. (1975) 'External control and regional development: the case of Scotland', *Environment and Planning*, A, 7.

Fothergill S. and Gudgin G. (1979) 'Regional employment change: a sub-regional explanation', *Progress in Planning*, **12** (3).
Foord J. (1980) 'Women's place – women's space: comment', *Area*, **12** (1).
Foord J. and Mackenzie S. (1981) 'Women and the city – towards a feminist analysis'. Unpublished Conference Paper, Social Administration Association.
French M. (1979) *The Woman's Room*. Sphere, London.
Friedan B. (1968) *The Feminine Mystique*. Penguin, Harmondsworth.
Garrett P.M. (1976) 'Some structural constraints on the agricultural activities of women: the Chilean hacienda'. *Research Paper No. 70*. Land Tenure Centre, University of Wisconsin.
Gasson R. (1982) 'Opportunities for women in agriculture', *Department of Environmental Studies and Countryside Planning, Occasional Paper*, No. 5, Wye College, London.
Greenhalgh C. (1977) 'Participation and Hours of Work for Married Women in Britain'. *L.S.E. Centre for Labour Economics Discussion Paper-No.13*.
Gregory J. (1982) 'Equal Pay and Sex discrimination: Why women are giving up the fight', *Feminist Review*, **10**.
Hagerstrand T. (1970) 'What about people in regional science?', *Papers of the Regional Science Association'*, **24**.
Hakim C. (1979) *Occupational Segregation. Research Paper No. 9. Dept. of Employment.* HMSO, London.
Hakim C. (1981) 'Job Segregation: trends in the 1970s', *Employment Gazette*. Dept of Employment **89** (12).
Hall, C. (1982) 'The butcher, the baker, the candlestick maker: the shop and the family in the Industrial Revolution', in Whitelegg E. *et al.* (eds), *The Changing Experience of Women*, Martin Robertson, Oxford.
Hall C. and Himmelweit S. (1981) *Public and private spheres*, Open University Course U221, Unit 8, Open University Press, Milton Keynes.
Hall J. and Perry N. (1974) 'Aspects of leisure in two industrial cities'. *Occasional Papers in Survey Research No. 5*. Social Science Research Council Survey Unit.
Hall P., Gracey H., Drewett R. and Thomas R. (1973) *The Containment of Urban England*. George Allen and Unwin, London.
Harre R. (1979) *Social being: a theory for social psychology*. Blackwell, Oxford.
Harry I.S. (1980) 'A study of the sexual division of labour by farm type: Women in Agriculture in Trinidad'. Unpublished M.Sc thesis, University of Calgary, Canada.
Hartmann H. (1981) 'The unhappy marriage of marxism and feminism: towards a more progressive union', in Sargent L. (ed), *Women and Revolution*. Pluto Press, London.
Harvey D. (1973) *Social Justice and the City*. Edward Arnold, London.
Harvey D. (1975) 'Class structure in a capitalist society and the theory of residential differentiation', in Peel R., Chisholm M. and Haggett P. (eds), *Processes in Physical and Human Geography*. Heinemann, London.
Harvey, D. (1978) 'Labour, capital and class struggle around the built environment in advanced capitalist societies' in Cox K.R. (ed), *Urbanization and Conflict in Market Societies*, Methuen, London.

Hayden D. (1980a) *The Grand Domestic Revolution*, MIT Press, Cambridge, Mass.
Hayden D. (1980b) 'What would a non-sexist city be like?', *Signs*, 5 supplement to issue 3, special issue on 'Women and the American city', 170-187.
Hayford A. (1974) 'The geography of women: an historical introduction', *Antipode*, **6**.
Haynes R. and Bentham C. (1982) 'The effects of accessibility on general practitioner's consultations, out-patient attendances and in-patient admissions in Norfolk, England', *Social Science and Medicine*, **16** (5).
Henshall J.D. (1981) 'Women and small scale farming in the Caribbean', in Horst O. (ed), *Papers in Latin American Geography in Honor of Lucia C. Harrison*. Conference of Latin American Geographers Special no. Publications 1, Muncie, Indiana.
Henshall Momsen J. (1980) 'Women in Canadian Geography', *Professional Geographer*, **32**, 3, 365-369.
Hibberd, C. and Hindle, J. (1981) 'Perceived areas of fear and activity patterns: a study of women undergraduates at Reading University'. Unpublished undergraduate project in Social and Behavioural Geography, Reading University.
Hicks C. (1982) 'Racism in nursing', *Nursing Times*, 5 and 12 May.
Hillman M. (1970) 'Mobility in New Towns'. Unpublished Ph.D thesis, University of Edinburgh.
Horst O. (1981) *Papers in Latin American Geography in Honor of Lucia C. Harrison*. Conference of Latin Americanist Geographers. Special Publication No. 1, Muncie, Indiana.
Hoyt, H. (1939) *The Structure and Growth of Residential Neighbourhoods in American Cities*, Federal Housing Administration, Washington, D.C.
Hudson R. (1980) 'Women and Work: A Study of Washington New Town'. *Occasional Papers (New Series) 16*. Dept of Geography, University of Durham.
Hudson R. (1982) 'Accumulation, spatial policies and the production of regional labour reserves: a study of Washington New Town', *Environment and Planning A*, **14**.
Huws U. (1983) *Your Job in the Eighties: A Woman's Guide to New Technology*. Pluto Press, London.
Jephcott P. (1962) *Married Women Working*. George Allen and Unwin, London.
Johnston R.J. and Brack E.V. (1983) 'Appointment and promotion in the academic labour market: a preliminary survey of British University Departments of Geography, 1933 – 1982', *Transactions of the Institute of British Geographers*, New Series, **8**.
Joshi H. and Owen S. (1981) 'Demographic Predictors of Women's Participation in Postwar Britain'. *Centre for Population Studies Working Paper*, No.81.
Keeble D. (1977) 'Spatial policy in Britain: regional or urban?', *Area*, 9.
Keeble D. (1981) 'Manufacturing dispersion and government policy in a declining industrial system: the U.K. case 1971 – 1976', in Rees J. *et al.* (eds), *Industrial Location and Regional Systems*. J. F. Bergin, New York.
Keller, S. (1981) *Building for Women*, Lexington Books, Lexington, Mass.
Kemener, B.J. (1965) 'A study of the relationship between the sex of the student and the assignment of marks by secondary school teachers'. Unpublished Ph.D thesis, Michigan State University, East Lansing, Michigan.
Klein V. (1965) *Britain's Married Women Workers*. Routledge and Kegan Paul, London.

Knox P. (1978) 'The intraurban ecology of primary medical care: patterns of accessibility and their policy implications', *Environment and Planning A*, **10**.
Knox, P. (1982) *Urban Social Geography*, Longmans, London.
Leeson J. and Gray J. (1978) *Women in Medicine*, Tavistock, London.
Lewis J. (1982) 'Changing Patterns of Gender Differentiation in Peterlee New Town 1948 – 1982'. *Working paper No. 2*. Dept of Geography, Queen Mary College, London University.
Ley D. (1983) *A Social Geography of the City*. Harper and Row, New York.
Low Pay Unit (1978) *The Part-time Trap*. LPU, London.
Low Pay Unit (1979) *The Hidden Army*. LPU, London.
Low Pay Unit (1980) *Minimum Wages for Women*. LPU, London.
McCrone G. (1969) *Regional Policy in Britain*. Unwin, University Books.
McDowell L. (1979) 'Women in British Geography', *Area*, **11**, 151–154.
McDowell, L. (1983) 'Towards an understanding of the gender division of urban space', *Environment and Planning D: Society and Space*, **1**, 59-72
Mackie L. and Patullo P. (1977) *Women at Work*. Tavistock Women's Studies, London.
Mackenzie S. (1980) 'Women and the reproduction of labour power in the industrial city', *Working Paper 23*, Urban and Regional Studies, University of Sussex, Brighton.
Mackenzie S. and Rose D. (1983) 'Industrial change, the domestic economy and home life', in Anderson J., Duncan S. and Judson R. (eds), *Redundant Spaces? Social Change and Industrial Decline in Cities and Regions*. Academic Press, London.
Mackintosh M. (1981) 'Gender and Economics: the sexual division of labour and the subordination of women', in Young K., Wolkowitz C. and McCullagh R. (eds), *Of Marriage and the Market: Women's Subordination in International Perspective*. CSE Books, London.
McNabb R. (1980) 'Segmented labour markets, female employment and poverty in Wales', in Rees G. and Rees T. (eds), *Poverty and Social Inequality in Wales*. Croom Helm, London.
McRobbie, A. (1980) 'Setting accounts with sub-cultures: a feminist critique', *Screen Education*, **34**, 37–49.
Marshall K. (1982) *Real Freedom: Woman's Liberation and Socialism*. Junius, London.
Massey D. (1979) 'In what sense a regional problem?', *Regional Studies*, **13**.
Massey D. (1982) 'The geography of industrial change', in Potter D. (ed), *Society and the Social Sciences*. Open University.
Massey D. (1983) 'Industrial restructuring as class restructuring: production decentralization and local uniqueness', *Regional Studies*, **17**, (2)
Massey D. and Meegan R. (1978) 'Industrial restructuring versus the cities', *Urban Studies*, **15** (3)
Matrix Book Group (1984) *Making Space: women in the man-made environment*. Pluto Press, London.
Miles N. (1982) 'The spatial decentralisation of manufacturing growth and the restructuring of the labour process'. *Discussion paper No.3 (New Series)*. Graduate School of Geography CSE.
Miller R. (1983) 'The Hoover in the garden: middle-class women and suburbanisation, 1850-1920', *Environment and Planning D: Society and Space*, **1**, 73-87.
Millet K. (1971) *Sexual politics*. Hart-Davis, London.

Monk J. (1981) 'Social Change and Sexual Differences in Puerto Rican Rural Migration', in Horst, O. (ed) *Papers in Latin American Geography in honor of Lucia C. Harrison*, Conference of Latin American Geographers, Special Publication No. 1, Muncie, Indiana.
Monk J. and Hanson S. (1982) 'On not excluding half of the human in human geography', *The Professional Geographer*, **34**, 11–23.
Muth R. (1969) *Cities and Housing*. Chicago University Press, Chicago.
Neal J. (1983) *Memoirs of a callous picket: working for the N.H.S.* Pluto, London
Nelson N. (1979) 'The sexual division of labour in the informal sector of a Nairobi squatter settlement', in Bromley R. and Gerry C. (eds), *Casual Work and Poverty in Third World Cities*. John Wiley, New York.
Oakley A. (1981) 'Interviewing women', in Roberts H. (ed) *Doing Feminist Research*. Routledge and Kegan Paul, London, pp. 30–61.
Oakley A. (1981) *Subject Women*. Martin Roberts, London
Oakley A. and Oakley R. (1979) 'Sexism in Official Statistics', in Irvine J., Miles I. and Evans J. (eds) *Demystifying Social Statistics*, Pluto Press, London.
Palm R. and Pred A. (1974) 'A time-geographic perspective on problems of inequality for women'. *Working Paper No. 236*. Institute of Urban and Regional Development, University of California, Berkeley.
Parker Morris Committee Report (1961) *Houses for Today and Tomorrow*, Ministry of Housing and Local Government, HMSO, London.
Paterson R. and Armstrong M. (1972) *An Employers Guide to Equal Pay*. Kogan Page Associates, London.
Perrons D. (1981) 'The role of Ireland in the new international division of labour', *Regional Studies*, **15** (2).
Phillips A. and Taylor B. (1980) 'Sex and Skill: notes towards a feminist economics', *Feminist Review*, No. 6.
Piercy M. (1978) *Woman on the Edge of Time*. Women's Press, London.
Piercy M. (1980) *Vida*. The Women's Press, London.
Planning and Women Group (1982) *Women and the Planned Environment: Conference Proceedings*, PAW, London.
Pollert A. (1981) *Girls, Wives, Factory Lives*. Macmillan Press, London.
Randall M. (1981) *Sandino's Daughters. Testimonies of women involved in the Nicaraguan fighting*. Zed Press, London.
Reith Committee Report (1946) *Final Report, New Towns Committee*, Cmnd 6876, HMSO, London.
Relph E. (1976) *Place and placelessness*. Pion, London.
Relph E. (1981) *Rational landscapes and humanistic geography*. Croom Helm, London.
Rengert A. (1981) 'Some socio-cultural aspects of rural outmigration in Latin America', in Horst O. (ed) *Papers in Latin American Geography in honor of Lucia C. Harrison*, Conference of Latin American Geographers, Special Publication No. 1, Muncie, Indiana.
Rex J. and Moore R. (1967) *Race, Community and Conflict*, Oxford University Press, Oxford.
Rimner L. and Popay J. (1982) 'The family at work: Special feature', *Employment Gazette*. Dept of Employment.
Roberts H. (ed) (1981) *Doing Feminist Research*. Routledge and Kegan Paul, London.
Roberts M. (1981) 'British Restaurants (1940-1946): from collectivism to

consumerism', paper presented at a conference on Women and Housing Policy, University of Kent, Canterbury, Kent.
Rogers B. (1980) *The Domestication of Women*. Kogan Page, London.
Rose D. (1981) 'Accumulation versus reproduction in the inner city: the recurrent crisis of London revisited', in Dear M. and Scott A. (eds) *Urbanisation and Urban Planning in a Capitalist Society*. Methuen, London, 339–381.
Rowbotham S. (1977) *Hidden from history: 300 years of women's oppression and the fight against it* (3rd edition). Pluto, London.
Rowbotham S., Segal L. and Wainwright H. (1979) *Beyond the fragments: feminism and the making of socialism*. Merlin, London.
Rubbo A. (1974) 'The spread of capitalism in rural Columbia: effects on poor women', in R. Reither (ed) *Toward an Anthropology of Women*. Monthly Review Press, New York.
Safa H.I. (1981) 'Runaway Shops and female employment: the search for cheap labour'. *Signs*, 7 (2).
Sando R.A. (1981) 'Doing the work of two generations: the Impact of depopulation on rural women in contemporary Taiwan'. *Paper presented at the International Geographical Union Commission on Rural Development International Symposium*: Fresno, California.
Sayer A. (1982) 'Explanation in Economic geography: abstraction versus generalisation', *Progress in Human Geography*, **6** (1).
Sayer A. (1983) 'Beyond the Fragments and the meaning of socialism'. Proceedings of a one-day conference in *Meeting of Minds*. Reading: IBG Women and Geography Study Group.
Seamon D. (1979) *A Geography of the Lifeworld*. Croom Helm, London.
Scott J. and Tilly L. (1975) 'Women's work and the family in nineteenth-century Europe', in Whitelegg E. *et al.* (eds) *The Changing Experience of Women*, Martin Robertson, Oxford.
Short J. (1980) *Urban Data Sources*, Butterworths, London.
Smith D.M. (1973) *Human Geography: a Welfare Approach*. Edward Arnold, London.
Smith D.M. (1982) 'Geographical perspectives on health and health care', in *Contemporary Perspectives on Health and Health Care*. Research Group QMC , Occasional Paper No. 20 , Department of Geography Queen Mary College, London.
Snell M., Glucklich P. and Porall M. (1981) 'Equal Pay and Opportunity: A study of the Implementation and Effects of the Equal Pay and Sex Discrimination Acts in 26 Organisations'. *Dept of Employment Research Paper*, No. 20.
Spender D. (1982) *Invisible women: the schooling scandal*. Writers and Readers Publishing Group, London.
Stanworth M. (1983) *Gender and Schooling*. Hutchinson, London.
Stavrakis O. and Marshall M.L. (1978) 'Women agriculture and development in the Maya Lowlands: Profit or Progress. The impact of commercial sugar growing in Belize', in Cowan A.B. (ed) *Proceedings of the International Conference on Women and Food*, Vol. 1. University of Arizona, Tuscon.
Stoddart D.R. (1983) 'IBG celebrates 150 years', *The Geographal Magazine*, January, 40–41.
Stoler A. (1977) 'Class Structure and Female Autonomy in Rural Java', *Signs*.

Taylor B. (1983) *Eve and the New Jerusalem*. Virago, London.
The Thornhill Neighbourhood Project (1982) *Ante-Natal Care – who benefits?*. Thornhill Neighbourhood Project, London.
Thrift N. (1976) *An Introduction to Time Geography*. Concepts and Techniques in Modern Geography (CATMOG), 13. Geo Abstracts Ltd, Norwich.
Tivers J. (1976) 'Constraints on spatial activity patterns: women with young children', *Occasional Paper*, **6**, Kings College, Department of Geography, University of London.
Tivers J. (1978) 'How the other half lives: the geographical study of women', *Area*, **10**, 302-306.
Tivers J. (1982) 'Weekday spatial activity patterns of women with young children'. Unpublished Ph.D thesis, University of London.
Tizard J., Moss P. and Perry J. (1976) *All Our Children*. Temple Smith/New Society.
Tolson A. (1977) *The limits of masculinity*. Tavistock Press, London.
Townsend P. (1982) 'A study of fear of attack on Campus and in Halls of Residence', Unpublished undergraduate project in Social Geography, Reading University.
Townsend P. and Davidson N. (1980) *Inequalities in Health: the Black Report*.
Tucker J. (1966) *Honorable Estates*, Victor Gollancz, London.
Tysoe M. (1982) 'Do colleges mark women down?' *New Society*, **62**, 9 December, 429-430
Ungerson C. (1971) *Moving Home*. Bell and Hyman, London.
Unwin R. (1901) *The Art of Building a Home*, Longman, Harlow, Essex.
Urry J. (1981) 'Localities, regions and social class', *International Journal of Urban and Regional Research*, 5 (4).
Walby S. (1983) 'Women's unemployment: some spatial and historical variations'. *Paper to SSRC conference on Urban Change and Conflict*.
Walker A. (1982) *Meridian*, The Women's Press, London.
Ward C. (1974) *Tenants Take Over*. Architectural Press, London.
Watt I. (1982) 'Occupational stratification and the sexual division of labour: Scotland since 1945', in Dickson T. (ed) *Capital and Class in Scotland*. John Donald, Edinburgh.
Werkerle G., Peterson R. and Morely D. (eds) (1980) *New Space for Women*. Westview, Boulder, CO.
West J. (1982) *Work, Women and the Labour Market*. Routledge and Kegan Paul, London.
Williamson L. (1982) 'Industrial restructuring, local class structure and female waged work on Merseyside', *Working Paper 32, Urban and Regional Studies*. University of Sussex.
Whitelegg E. *et al.* (eds) (1982) *The Changing experience of women* Martin Robertson, Oxford.
Whitelegg J. (1982) *Inequalities in Health Care: Problems of access and provision*, Straw Barnes, Nottinghamshire.
Young K. (1978) 'Modes of Appropriation and the Sexual Division of Labour: A case study from Oaxaca Mexico', in Kuhn A. and Wolpe A. M. (eds) *Feminism and Materialism: Women and Modes of Production*. Routledge and Kegan Paul, London.
Zelinsky W., Monk J. and Hanson S. (1982) 'Women and geography: a review and prospectus, *Progress in Human Geography*, **6** (3), 317–366.

Index

FSC
www.fsc.org
MIX
Papier | Fördert gute Waldnutzung
FSC® C083411

Kolibri 360 GmbH
Ferdinand-Jühlke-Straße 7
99095 Erfurt, Deutschland
produktsicherheit@kolibri360.de